FM 17-79

DEPARTMENT OF THE ARMY FIELD MANUAL

TANK 90-MM GUN M48

FIELD MANUAL

BY DEPARTMENT OF THE ARMY • OCTOBER 1955

DISCLAIMER:

This document is a reproduction of a text first published by the Department of the Army, Washington DC. All source material contained herein has been approved for public release and unlimited distribution by an agency of the US Government. Any US Government markings in this reproduction that indicate limited distribution or classified material have been superseded by downgrading instructions promulgated by an agency of the US government after the original publication of the document No US government agency is associated with the publication of this reproduction. This manual is sold for historic research purposes only, as an entertainment. It contains obsolete information and is not intended to be used as part of an actual training program. No book can substitute for proper training by an authorized instructor.

FIELD MANUAL ⎰   DEPARTMENT OF THE ARMY
No. 17-79   ⎱   WASHINGTON 25, D. C.. *13 October 1955*

TANK, 90-MM GUN, M48

---

*This manual supersedes TC 21, 6 October 1953; and so much of chapter 18, FM 17-12, 30 November 1950, as pertains to Tank, 90-mm Gun, M48.

CHAPTER 1

INTRODUCTION

1. Purpose and Scope

a. The purpose of this manual is—

(1) To give the general characteristics of the Tank, 90–mm Gun, M48.

(2) To explain in detail the armament, the turret controls, the fire-control instruments, and the auxiliary fire-control equipment in the Tank, 90–mm Gun, M48.

(3) To provide a guide for personnel of armored units in learning and teaching the fire commands, firing duties, crew drill, and service of the piece for the M48 tank.

b. This manual covers materiel, crew drill, service of the piece, conduct of fire, and the tank gunnery qualification course for the Tank, 90–mm Gun, M48.

CHAPTER 2

MATERIEL

Section I. GENERAL CHARACTERISTICS

2. Description

a. The Tank, 90–mm Gun, M48 (fig. 1), is an armored, full-track, combat vehicle of the medium-gun tank class. It is divided into three compartments: the fighting compartment in the turret for the tank commander, gunner, and loader; the driving compartment in the front of the hull for the driver; and the engine compartment in the rear of the hull for the engine and transmission. A rotatable dome commander's cupola is mounted to the right side of the turret roof.

b. The M48 tank shown in figure 2 is identical to the one shown in figure 1 with the exception that it is provided with the caliber .50 Machinegun Mount, Stock No. 7364875, in lieu of the tank commander's cupola.

c. The tank has an eliptical hull and turret to provide maximum ballistic protection against antitank projectiles. Improved ammunition stowage and a more efficient fire-control system have increased the firepower, accuracy, and speed of engaging targets over those of previous tanks.

3. Data, Tank, 90-mm Gun, M48

Crew	Four
Armament	One Gun, 90-mm, M41; one Machinegun, Caliber .50; one Machinegun, Caliber .30.
Communication system	Radio and interphone.
Weight, full equipped (approximately)	49½ tons.
Length, overall (gun in traveling position)	24 feet, 1½ inches.
Width	12 feet.
Height, overall	10 feet, 7⅝ inches.
Ground clearance	18 inches.
Ground pressure	10.2 pounds per square inch.
Electrical system	24 volts.
Maximum grade-ascending ability	60 percent.
Maximum fording depth	4 feet.
Minimum turning circle	Pivot.

Figure 1. Tank, 90-mm Gun, M48.

5

Figure 2. Tank, 90-mm Gun, M48.

6

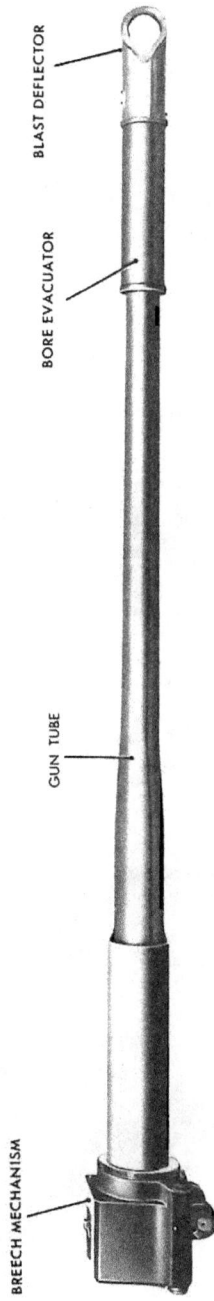

BLAST DEFLECTOR

BORE EVACUATOR

GUN TUBE

BREECH MECHANISM

Figure 3. Gun, 90-mm, M41.

7

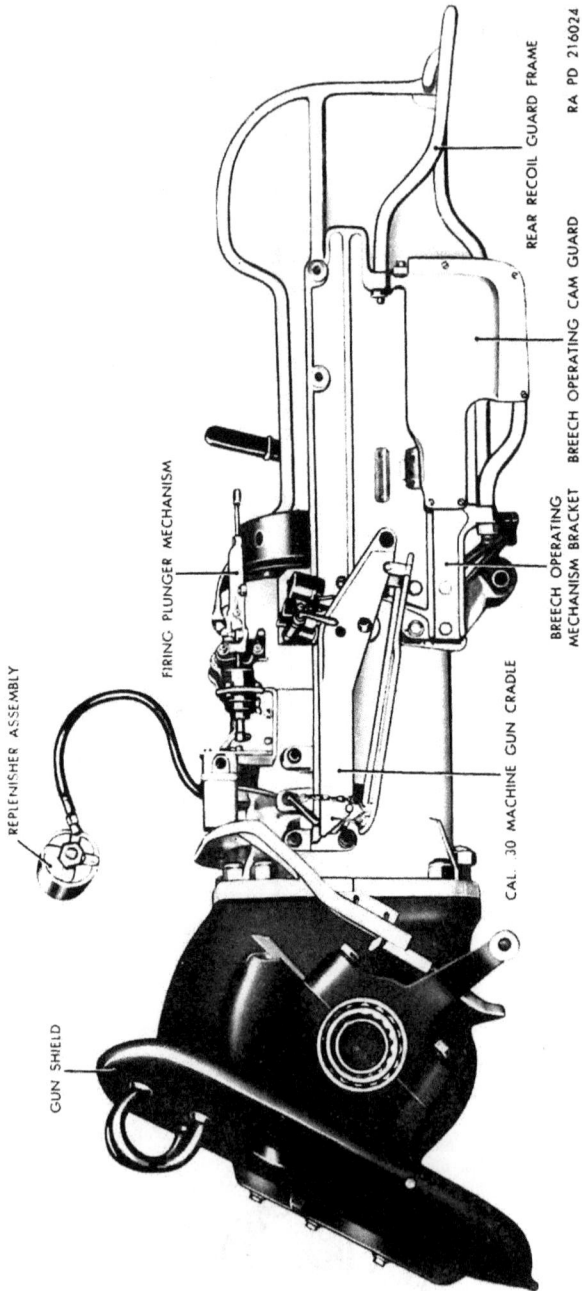

REPLENISHER ASSEMBLY

GUN SHIELD

FIRING PLUNGER MECHANISM

REAR RECOIL GUARD FRAME

RA PD 216024

BREECH OPERATING CAM GUARD

BREECH OPERATING
MECHANISM BRACKET

CAL. 30 MACHINE GUN CRADLE

Figure 4. Gun in mount, left side.

Section II. GUN, 90-MM, M41

4. General

a. The Gun, 90-mm, M41, is designed primarily for the medium-gun tank. It consists of four major parts: the tube, bore evacuator, blast deflector, and breech mechanism (fig. 3).

b. The gun is mounted in the Combination Gun Mount, T148, which consists of a gun shield and a cradle (fig. 4). The gun mount supports the gun on trunnion bearings and provides attachments for the breech operating and firing mechanisms, the coaxial machinegun mount, and the recoil guards. The gun mount is designed to provide for a quick change of the gun tube.

c. The tube is formed in one piece, threaded at the breech end for attachment of the breech mechanism and at the muzzle end for assembly of the bore evacuator and attachment of the blast deflector (fig. 5). The bore of the tube is rifled with grooves having a uniform right-hand twist and is chrome-plated over the full length of the rifling.

d. The bore evacuator (fig. 5) is formed by a thin-walled cylinder fitted around the forward end of the tube to form an evacuator chamber. Eight holes, drilled into the bore of the tube and slanted at an angle of 30° toward the muzzle, connect the evacuator chamber to the bore. The bore evacuator removes residual gases, reducing crew discomfort caused by these gases escaping into the fighting compartment.

e. The blast deflector (fig. 5) is a one-piece steel casting which is screwed onto the gun tube forward of the bore evacuator. Its primary function is to reduce obscuration of the target by muzzle blast.

f. The breech mechanism consists of the breech ring, the breechblock with its component parts, and the breech operating mechanism (figs. 6, 7, and 8).

5. Data, Gun, 90-mm, M41

Caliber	90 millimeters.
Length of bore	14 feet, 9 inches.
Type of breechblock	Vertical sliding wedge.
Maximum chamber pressure	47,000 pounds per square inch.
Type of recoil mechanism	Concentric, hydrospring.
Normal recoil	12 inches.
Maximum recoil	13.5 inches.
Maximum elevation	20° (355 mils).
Maximum depression	9° (160 mils).
Weight of gun complete	2370 pounds.
Weight of tube only	1582 pounds.
Ammunition and approximate velocity:	
HE and WP	2,400 feet per second.
AP	3,050 feet per second.
HVAP	4,050 feet per second.
HVAP–DS	4,100 feet per second.
HEAT	2,800 feet per second.

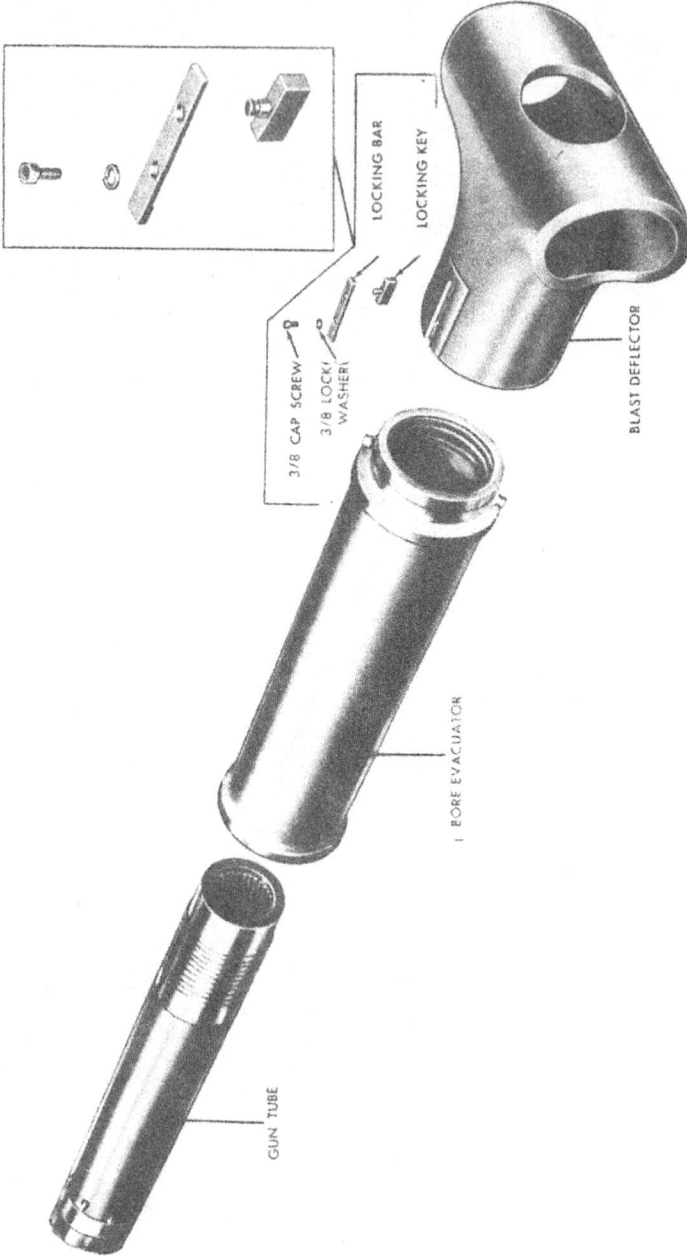

Figure 5. Gun tube, bore evacuator, and blast deflector.

GUN TUBE

BORE EVACUATOR

3/8 CAP SCREW

3/8 LOCK
WASHER

LOCKING BAR

LOCKING KEY

BLAST DEFLECTOR

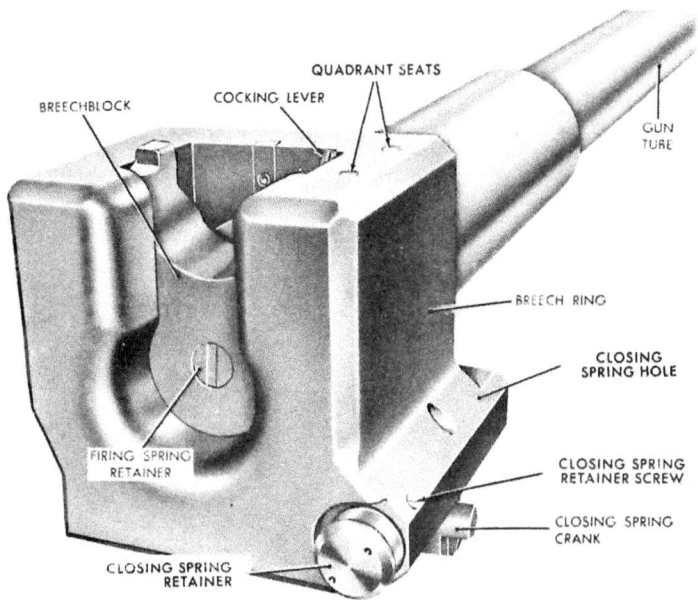

QUADRANT SEATS
COCKING LEVER
BREECHBLOCK
GUN TUBE
BREECH RING
CLOSING SPRING HOLE
FIRING SPRING RETAINER
CLOSING SPRING RETAINER SCREW
CLOSING SPRING CRANK
CLOSING SPRING RETAINER

Figure 6. Breech mechanism.

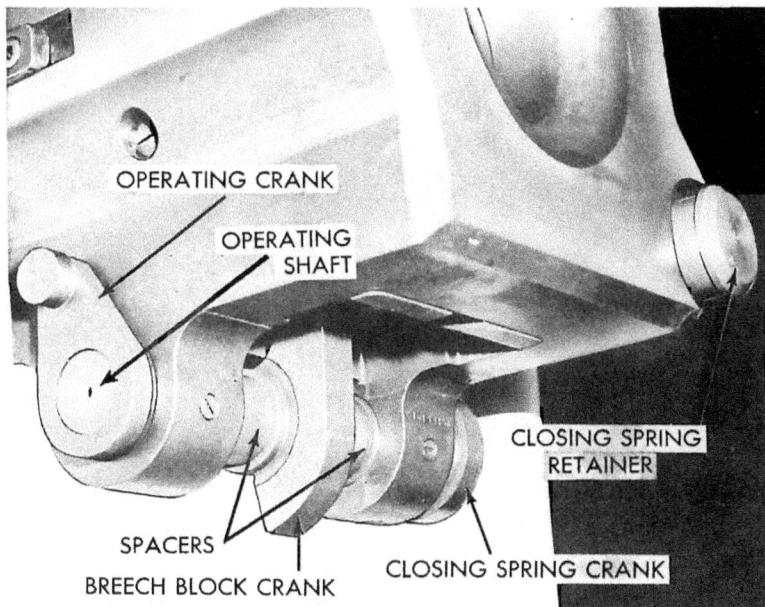

OPERATING CRANK
OPERATING SHAFT
CLOSING SPRING RETAINER
SPACERS
BREECH BLOCK CRANK
CLOSING SPRING CRANK

Figure 7. Breech operating mechanism.

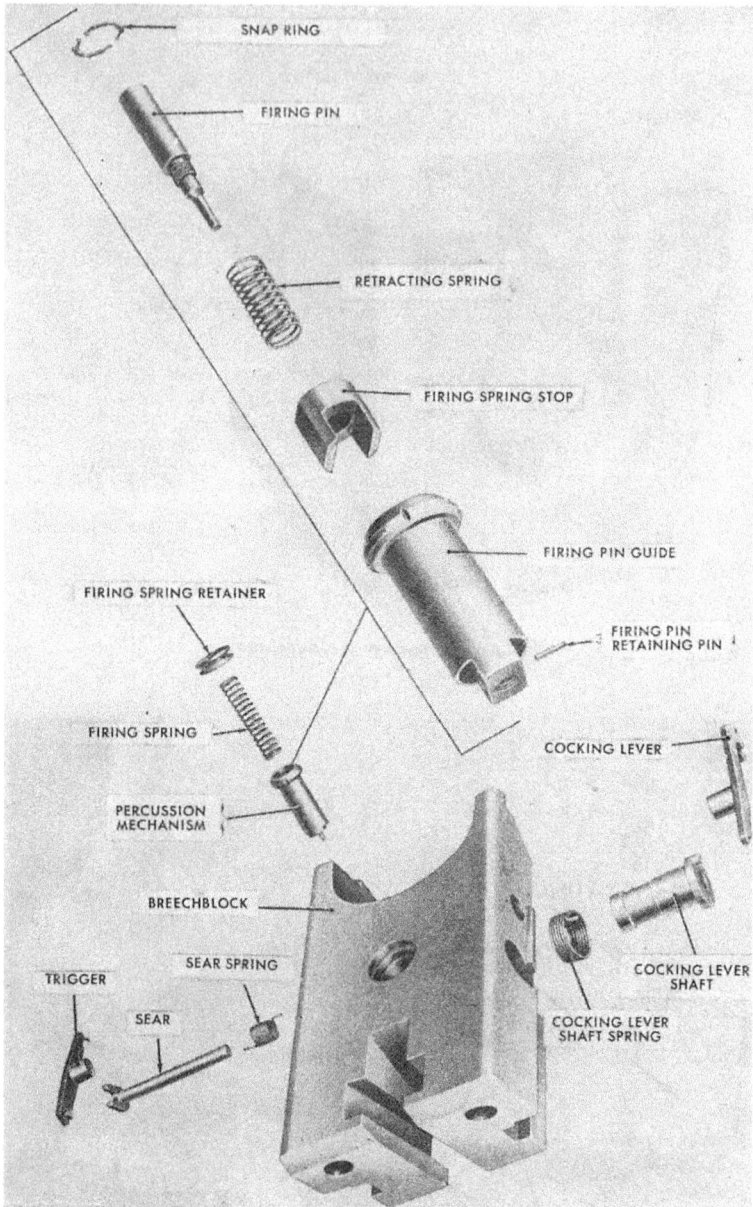

SNAP RING

FIRING PIN

RETRACTING SPRING

FIRING SPRING STOP

FIRING PIN GUIDE

FIRING SPRING RETAINER

FIRING PIN
RETAINING PIN

FIRING SPRING

COCKING LEVER

PERCUSSION
MECHANISM

BREECHBLOCK

COCKING LEVER
SHAFT

TRIGGER

SEAR SPRING

SEAR

COCKING LEVER
SHAFT SPRING

Figure 8. Cocking and firing mechanism, exploded view.

12

6. Disassembly and Assembly, Gun 90-mm, M41

a. General. The disassembly and assembly procedures outlined herein are intended as guides. Because of the weight of the breech-block, extreme care must be taken to avoid injury to personnel during disassembly and assembly of the gun. At least two persons working together should perform disassembly and assembly of the weapon.

b. Disassembly.

(1) Open the breech slightly and inspect the chamber.

(2) Operate the firing trigger or hand firing lever to release the compression of the firing spring.

(3) Remove the firing spring retainer by pressing it inward and rotating it 90° until the slot in the retainer is horizontal.

(4) Remove the firing spring. Hold one hand over the percussion mechanism well, and pull the cocking lever to the rear. This moves the percussion mechanism to the rear past the sear. Grasp the rear end of the percussion mechanism and remove it from the percussion mechanism well.

> *Note.* Disassembly of the percussion mechanism should be performed only for inspection or when a malfunction or defective part makes it necessary. Procedure is as follows: Press the firing spring stop into the firing pin guide, and drift out the firing pin retaining pin. Unscrew the firing pin from the guide, releasing the retracting spring and the firing spring stop (fig. 8).

(5) Remove the operating handle retaining bolt from the hub of the operating handle. Insert a drift or small-shank screwdriver through the hub, and depress the operating shaft plunger with one hand while removing the operating shaft retainer with the other hand. Replace and tighten the bolt in the hub of the operating handle to secure it to the right recoil guard.

(6) Insert the plug of the breechblock-removing tool in the percussion mechanism well so that the long flange of the tool rests in the loading notch of the breech ring.

(7) Rotate the operating handle to the rear until the weight of the breechblock is on the removing tool and no longer on the operating shaft. Push the operating shaft to the left and remove it, catching the spacers as the operating shaft is being withdrawn. Relatch the operating handle.

> *Note.* If the operating shaft binds, move the operating handle up and down slightly until the shaft is free.

(8) Screw the eyebolt into the top of the breechblock, and depress the muzzle of the gun fully. Attach a rope (or **S**-hook) to the eyebolt and tie it tightly to the eye welded on the inside of the turret roof. Elevate the muzzle slightly to place the weight of the breechblock on the rope, and remove the breechblock-removing tool.

(9) Elevate the muzzle of the gun until the bottom of the breech-block clears the loading notch of the breech ring, and remove the breechblock crank, taking care not to drop the crossheads.

(10) Elevate the muzzle of the gun until the bottom of the breech-block is clear of the breech ring. Push the breechblock forward until it is in line with the forward edge of the breech ring. Depress the muzzle and allow the breechblock to rest flat on top of the breech ring, being careful not to damage the quadrant seats or firing linkage.

(11) Remove the extractors from the breech ring.

(12) Remove the cocking lever, cocking lever shaft, and cocking lever shaft spring from the right side of the breechblock.

(13) Remove the trigger, sear, and sear spring from the left side of the breechblock.

(14) The closing spring will not be removed during field disassembly. However, the closing spring may be removed using the following steps:

(a) After field disassembly, replace the operating shaft through the operating crank and closing spring crank; unlatch the operating handle and remove the operating handle stop from the right recoil guard.

(b) Allow the operating handle to rotate forward until the closing spring is fully expanded.

(c) Remove the closing spring retainer screw; unscrew the closing spring retainer and remove the closing spring.

c. *Assembly.*

(1) If the closing spring has been removed, proceed as follows:

(a) Replace the closing spring and closing spring retainer. Aline the second adusting hole of the retainer so that it will be engaged by the closing spring retainer screw.

(b) Pull the operating handle to the rear far enough to allow replacement of the operating handle stop.

(c) Replace the operating handle stop, and latch the operating handle.

(d) Remove the operating shaft.

(2) Replace the sear spring, sear, and trigger into the left side of the breechblock.

(3) Replace the cocking lever shaft spring, cocking lever shaft, and cocking lever into the right side of the breechblock.

Note. The outer end of the cocking lever shaft spring fits into the hole nearest the cocking lever lug on the cocking lever shaft. If the spring becomes weak, place the end in the hole away from the cocking lug.

(4) Replace the extractors into the breech ring.

(5) Elevate the muzzle, and guide the bottom of the breechblock into alinement with the breechblock recess. Depress the

14

muzzle and at the same time guide the breechblock into the breech ring, making sure the extractor lips are forward. Place the breechblock crank, with the crossheads, in the inclined **T**-slot of the breechblock through the loading notch of the breech ring.

   Caution: Trip the extractors from underneath the breech ring.

(6) Insert the breechblock-removing tool into the percussion mechanism well. Continue depressing the muzzle until the weight of the breechblock is on the breechblock-removing tool.

(7) With the double spline of the operating shaft at 3 o'clock and the operating crank lug at 11 o'clock, insert the operating shaft through the operating crank, left spacer, breechblock crank, right spacer, and closing spring crank. To aline the closing spring crank, it may be necessary to pull the operating handle slightly to the rear. Make sure the breechblock crank stop is to the rear.

(8) Place the operating shaft retainer in the slot of the operating shaft after latching the operating handle.

(9) Remove the rope (S-hook), eyebolt, and breechblock-removing tool.

(10) Replace the percussion mechanism, pulling on the hand firing lever so that the percussion mechanism can go fully forward.

(11) Replace the firing spring and firing spring retainer.

(12) Cock and actuate the firing mechanism.

   Note: Never trip the extractors with the fingers. Use the ramming and extracting tool, the base of an empty round, or a block of wood.

7. Functioning of Gun, 90-mm, M41

   a. Manual Opening of Breech. Grasp the operating handle, unlatch it, and pull it to the rear and down. As the operating handle is rotated to the rear, a lug on the operating handle hub contacts a similar lug on the closing spring crank. The closing sprink crank, in turn, rotates to the rear. Since it is splined to the operating shaft, it rotates the shaft, the breechblock crank, and the operating crank. The gear sector on the closing spring crank moves the closing spring piston to the rear, compressing the closing spring between the forward end of the closing spring piston and the closing spring retainer at the rear. The cross heads of the breechblock crank, riding in the inclined **T**-slot of the breechblock, move the breechblock to the open position. As the breech opens, the trunnions of the extractors, moving in the curved extractor grooves of the breechblock, are forced forward; and the flat surfaces of the trunnions are positioned directly above the trunnion seats of the breechblock. The closing spring expands slightly, moving the breechblock upward until the trunnion seats contact the flat surfaces of the trunnions. This locks the breechblock in the open position. The extractor plunger springs expand to insure the locking of the breech in the open position.

b. Cocking. As the breech opens and the upper arm of the cocking lever is cammed to the rear by the camming surface of the breech ring, the lower arm cams the outer lug of the cocking lever shaft forward. When the shaft is rotated, the inner lug contacts the collar of the percussion mechanism and moves it to the rear, compressing the firing spring between the firing spring stop and the firing spring retainer. The rearward movement of the percussion mechanism rotates the sear, by engaging the flat portion in the center of the sear. This winds the sear spring. When the collar of the percussion mechanism is far enough to the rear to clear the sear, the sear spring unwinds, rotating the sear back to its normal position, causing the sear to hold the percussion mechanism in the cocked position.

c. Automatic Closing of Breech (Loading). Automatic closing of the breech occurs when a round of ammunition is loaded into the chamber. The rim of the cartridge case contacts the lips of the extractors and pushes them forward, forcing the trunnions of the extractors off the trunnion seats of the breechblock and into the curved extractor grooves. The breechblock moves upward as the expanding closing spring forces the closing spring piston to rotate the closing spring crank, operating shaft, and breechblock crank. The crossheads of the breechblock crank, riding in the inclined **T**-slot of the breechblock, move the breechblock to the closed position.

d. Firing. As the trigger plunger is depressed, it contacts the upper arm of the trigger and pushes it to the rear. The lower arm of the trigger moves forward. Because it is in contact with the sear, it causes the sear to rotate, winding the sear spring. Rotation of the sear releases the percussion mechanism, allowing the firing spring to expand and to move the percussion mechanism forward in the breechblock. The sear spring unwinds, rotating the sear back to its normal position. As the percussion mechanism moves forward, the firing spring stop contacts the inner face of the breechblock and halts the expansion of the firing spring. The firing pin and guide continue forward under inertia. The firing pin strikes the primer of the round through the firing pin well in the forward face of the breechblock. This final forward movement compresses the retracting spring between the firing spring stop in front and the snap ring at the rear of the firing pin. After the firing pin strikes the primer, the retracting spring expands, withdrawing the firing pin from the primer and into the firing pin well.

e. Automatic Opening of Breech.   When the gun recoils after firing, the lug on the operating crank moves the operating cam on the left side of the gun mount away from the gun, compressing the operating cam return spring.   As the lug clears the cam, the return spring expands, moving the cam back to its normal position.   During counterrecoil, the operating crank lug strikes the operating cam and is rotated to the rear.   Because the operating crank is splined to the operating shaft, the operating shaft and breechblock crank rotate, moving the breechblock to the open position.   At the same time, the closing spring crank, being splined to the operating shaft, moves the closing spring piston to the rear, compressing the closing spring.   As the breechblock reaches the open position, the extractor trunnions, moving in the curved extractor grooves of the breechblock, are forced forward, and the flat surfaces of the trunnions are positioned directly above the trunnion seats of the breechblock.   The closing spring expands slightly, moving the breechblock upward until the trunnion seats contact the flat surfaces of the trunnions.   The extractor plunger springs expand, holding the extractor trunnions forward to insure the locking of the breechblock in the open position.

f. Extraction and Ejection.   As the breechblock nears the fully opened position and has cleared the rear of the cartridge case, the extractor lips are rotated to the rear as the extractor trunnions move forward in the curved extractor grooves.   The lips of the extractors, being in front of the rim of the cartridge case, extract the case from the chamber and eject it from the breech ring.

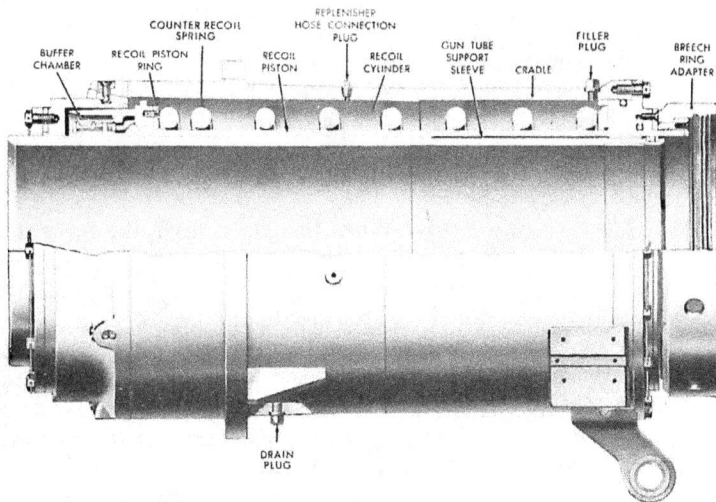

Figure 9.   Recoil cylinder assembly.

17

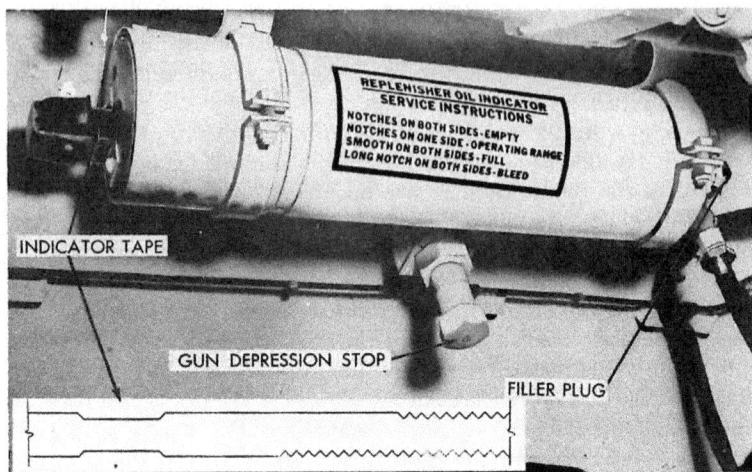

Figure 10. Replenisher assembly.

8. Recoil Mechanism of Gun, 90-mm, M41

a. General. Major components of the concentric hydrospring type recoil mechanism are the recoil cylinder assembly and the replenisher (figs. 9 and 10).

b. Recoil Cylinder Assembly.

(1) *Description.* The recoil mechanism operates to bring the gun to rest at the end of a normal recoil distance of 12 inches and to return the gun to battery with a minimum of shock. It is composed of the cradle, gun support sleeve, recoil piston, and counterrecoil spring. The recoil cylinder is formed by the inner surfaces of the cradle and the recoil piston. The counterrecoil spring is coiled around the recoil piston between the piston ring and the rear of the cradle. When ready for operation, the recoil cylinder is completely filled with recoil oil.

(2) *Action during recoil.* When the gun is fired, the force that propels the projectile forward acts upon the gun and drives the tube and breech ring rearward. During this action the recoil piston, which is attached to the breech ring by the breech ring adapter, moves to the rear with the breech ring. Movement of the piston to the rear forces hydraulic oil from the rear to the front of the piston ring, through the space between the ring and cylinder. The taper of the cylinder wall increasingly restricts the flow of oil as the piston moves to the rear and brings the gun to rest at the end of recoil. The counterrecoil spring is compressed during the recoil movement.

18

(3) *Action during counterrecoil.* At the end of recoil, the compressed counterrecoil spring expands and moves the gun forward. Three inches from the end of counterrecoil, the recoil piston enters the buffer chamber in the front of the recoil cylinder. The movement of oil out of the buffer chamber is thereby restricted, and the cushioning effect of the restricted flow eases the gun into battery without undue shock.

c. *Replenisher Assembly.*

(1) The replenisher assembly (fig. 10) consists of the replenisher cylinder, the replenisher piston, the replenisher piston spring, the indicator assembly, a spring-loaded ball valve, and a hose connection. The replenisher is connected to the recoil cylinder by a flexible hose.

(2) During firing, the heat generated causes expansion of the oil. As the oil expands. it passes through the flexible hose into the replenisher and applies pressure against the replenisher piston. The piston moves, compressing the replenisher piston spring and causing the indicator tape to wind around the screw in the indicator assembly. The indicator is a steel tape which is saw-toothed on both edges at one end, has one saw-toothed edge and one smooth edge in the center, and is smooth on both edges near the other end. It terminates with a long notch on each side. The indicator is read as follows: When both saw-toothed edges are exposed, the system is low on oil and should be refilled; when one saw-toothed edge and one smooth edge are exposed, the proper amount of oil is in the system; when both smooth edges are exposed, there is an excess of oil, which should be removed. When the gun cools after firing, the oil contracts and the compressed replenisher piston spring expands, forcing the oil back into the recoil cylinder so that the cylinder is full at all times.

d. *Checking, Draining, and Filling.*

(1) *Before firing.* The recoil system is checked by the use of the indicator tape on the replenisher. This check is performed while the recoil system is cool.

(2) *Draining.* If there is too much oil in the system, remove the filler plug on the replenisher, hold a rag under the filler plug hole, and allow the necessary amount of oil to drain onto the rag. To drain the excess oil, it is necessary to push in on the spring-loaded ball valve in the filler plug hole with the nozzle of the filler gun.

> *Note:* Use gradual pressure on the ball valve to control the amount of oil flowing from the replenisher. Check the indicator tape frequently to avoid draining too much oil.

(3) *Filling.* If the amount of oil is low, remove the filler plug from the replenisher and the nozzle from the filler gun. Fill the gun with the proper oil, screw the filler gun hose loosely into the filler plug hole, push on the plunger to force the air out, screw the hose tight, and force the oil into the replenisher. Repeat this procedure until the proper amount of oil is indicated.

(4) *During firing.* The system should be checked periodically for leaks and for excessive oil due to overheating. After several rounds have been fired, the indicator tape may show both edges smooth as a result of recoil oil expansion. It is not necessary to drain oil from the replenisher when this occurs. However, if the long notches are exposed, the crew should drain oil until the two smooth edges are exposed on the indicator tape.

e. Bleeding. If the length of gun recoil is excessive and the replenisher indicator shows the correct amount of oil, there may be air in the recoil system. To correct this, the turret mechanic should bleed the recoil system.

9. Malfunctions, Gun, 90-mm, M41

a. General. A malfunction is an unintentional cessation of fire caused by a failure of some part of the gun, the mount, or the ammunition. Malfunctions are divided into four general classes: failure to load, premature firing, failure to fire, and failure to extract and eject. The malfunctions discussed here are the most common ones but do not necessarily include all the malfunctions which may occur.

b. Failure To Load. Breech will not fully close.

Cause	Correction
Dirty, bulged, or dented round of ammunition.	Remove round and load another.
Dirty chamber	Remove round and clean chamber.
Gun out of battery (obstruction between gun and mount or too much recoil oil).	Remove obstruction or drain excess oil from replenisher.
Improper adjustment of closing spring	Remove closing spring retainer screw, tighten retainer until the next hole in the retainer is in line with the screw hole, and replace the screw.

c. Premature Firing. Gun fires as breech closes.

Cause	Correction
Weak sear spring, causing sear to release percussion mechanism as breechblock closes.	Replace defective part.

20

d. Failure To Fire.

Cause	Correction
Dirt, wax, or excess grease in percussion mechanism well, preventing free movement of percussion mechanism.	Clean percussion mechanism and percussion mechanism well.
Worn or broken cocking lever, cocking lever shaft, or cocking lever shaft spring.	Replace defective part.
Worn or broken firing pin, firing spring, sear, or sear spring.	Replace defective part.
Defective primer in round of ammunition.	Remove round, load another round, and fire.

e. Failure To Extract and Eject.

Cause	Correction
Defective rim of cartridge case causes extractor lips to pull through rim, leaving case in chamber.	Pry or ram case out of chamber.
Broken or missing extractor plungers and springs permit the trunnions to slip off trunnion seats of breechblock, allowing breechblock to jam cartridge case in breech ring.	Replace missing or defective part.
Weak, broken, rusty, missing, or grease-packed operating cam return spring causes breech to fail to open on counterrecoil.	Replace or clean defective part.

10. Care, Cleaning, and Lubrication—Gun, 90-mm, M41

a. General.

(1) It is vitally important to keep all materiel in proper condition for immediate service. Tools, accessories, and lubricating, cleaning, and preserving materials are provided for this purpose.

(2) Proper lubrication at specified intervals is essential to the care and preservation of the materiel. For instructions, refer to LO 9–7012 and TM 9–7012.

(3) Protective covers for the gun and mount should be installed when the gun is not in service. If the materiel is not to be used for a considerable length of time, all of the exposed, unpainted surfaces should be cleaned with rifle bore cleaner, dried thoroughly, and covered with a coat of rust preventive compound.

(4) Clean all disassembled parts thoroughly before lubricating and assembling them. Tools and accessories, as well as the materiel, should be kept clean, free from rust, and protected with preservative oil as required.

(5) The quadrant seats on the breech ring must be protected. Do not place tools or other articles on them.

b. Before Firing.

(1) *Tube.* Check the bore and chamber for dirt and obstructions. Clean as necessary, but do not lubricate.

(2) *Breech mechanism.* Examine the breech mechanism for proper functioning and presence of corrosion. Clean and lubricate.

c. During Firing. Be alert for any malfunction. Lubricate as necessary.

d. After Firing.

(1) *Tube.* After firing and for three consecutive days thereafter, clean the bore thoroughly with rifle bore cleaner, making sure that all surfaces in the bore and chamber are well coated. Wipe the tube dry each time before applying additional bore cleaner. After the fourth cleaning, dry and oil with the prescribed lubricant unless the gun will be fired in the next 24 hours.

(2) *Breech mechanism.* Disassemble, clean with bore cleaner, wipe dry, lubricate, and assemble. Check for proper functioning and condition of parts.

(3) *Bore evacuator.* The bore evacuator should be cleaned each time the gun is cleaned. More frequent cleaning may be necessary if the residual gases from the firing of ammunition are not cleared from the gun tube by the bore evacuator. To clean, remove the blast deflector and the evacuator can. Clean the jets in the gun tube by inserting a piece of soft wire through each of them. Clean the exterior of the tube and the interior of the evacuator can with bore cleaner. Coat the cleaned parts with the prescribed preservative, and reassemble them.

Caution: Be careful not to damage the sealing lips in removing and replacing the bore evacuator.

Section III. TANK MACHINEGUNS

11. General

a. The firepower of the caliber .50 and caliber .30 machineguns greatly increases the shock effect of the M48 tank.

b. This section describes and illustrates the mounts for these machineguns and furnishes essential information for the crew to mount and operate these weapons.

c. M48 tanks may be equipped either with the Cupola Mount, M30, which mounts the Machinegun, Caliber .50, M2E1, Fixed (fig. 1), or with the Mount, Stock No. 7364875, which mounts the Machinegun, Caliber .50, M2, HB, Flexible (fig. 2). The caliber .30 coaxial machinegun may be either the M1919A4E1 or the M37.

22

12. Combination Gun Mount, T148

a. General. The Combination Gun Mount, T148, mounts a caliber .30 machinegun (M1919A4E1 or M37) coaxially with the 90-mm gun (fig. 4). The machinegun mount is secured to the tank-gun cradle. When the main armament is elevated and traversed, the machinegun is moved correspondingly.

b. Machinegun Mount. In the M48 tank, the coaxial machinegun is located on the left side of the T148 combination gun mount. Two bracket assemblies, front and rear, hold the machinegun when mounted.

c. Component Parts. The component parts of the coaxial machinegun mount are—

(1) Front mount bracket.
(2) Machinegun front locking pin.
(3) Coaxial machinegun cradle.
(4) Elevating and traversing mechanism.
(5) Machinegun rear locking pin.

Note: Check bolts on front mounting bracket frequently for tightness.

d. Installation of Coaxial Machinegun.

(1) Pull out the machinegun front and rear locking pins.
(2) Insert the muzzle of the machinegun into the hole in the gun shield.
(3) Set the machinegun into the front mounting bracket, aline the front mounting holes, and insert the front locking pin.
(4) Pull up the elevating and traversing mechanism and aline it with the rear mounting holes of the machinegun.
(5) Insert the rear locking pin, and the machinegun will be secured in the coaxial gun mount.

e. Adjustment of Firing Solenoid (fig. 11). Headspace must be adjusted prior to adjustment of the firing solenoid.

(1) The adjusting screw is located at the bottom of the solenoid support, between the solenoid and the machinegun elevating bracket. The lock screw is located on the machinegun elevating bracket.
(2) Loosen the two jam nuts with a $7/16$-inch open-end wrench. Loosen the lock screw with a screwdriver.
(3) Using a screwdriver, turn the adjusting screw clockwise to move the solenoid plunger away from the trigger of the machinegun or counterclockwise to move the solenoid plunger closer to the trigger of the machinegun.
(4) Adjustment is correct when there is $1/32$-inch clearance between the solenoid plunger and the trigger of the machinegun with the weapon cocked.

23

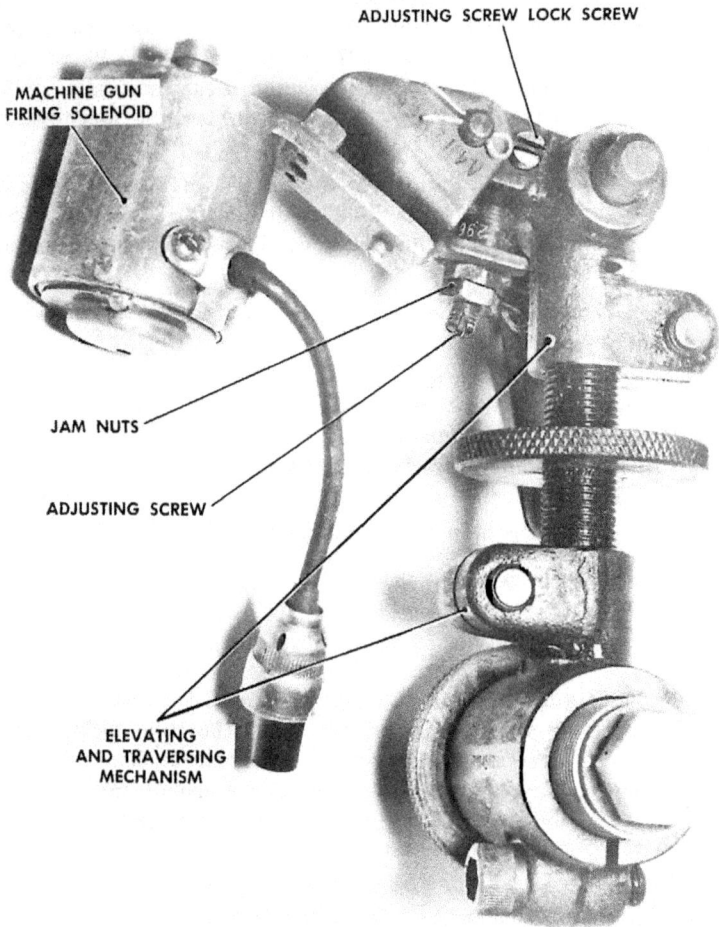

MACHINE GUN
FIRING SOLENOID

ADJUSTING SCREW LOCK SCREW

JAM NUTS

ADJUSTING SCREW

ELEVATING
AND TRAVERSING
MECHANISM

Figure 11. Solenoid and elevating bracket.

(5) When the adjustment is completed, tighten the lock screw and the two jam nuts on the solenoid adjusting screw. Check for final adjustment by holding the electric firing trigger while manually cocking the weapon several times to insure that the firing pin will be released as the bolt moves into battery.

f. Loading and Firing the Coaxial Machinegun.

(1) Load the ammunition tray, and run the belt into the feedway of the machinegun, double link first, until the first round is secured by the belt-holding pawl.

(2) To half-load the machinegun, pull the retracting handle to the rear and release it. Repeat the operation to full-load.

(3) To fire the gun electrically—

    (*a*) Turn on the caliber .30 machinegun firing switch.

    (*b*) Press any of the electric firing triggers.

g. Boresighting Coaxial Machinegun. The coaxial machinegun must be alined with the main armament. The back plate and bolt groups must be removed to boresight.

(1) To adjust the machinegun for elevation—

    (*a*) Loosen the two upper socket-head cap screws.

    (*b*) Turn the elevating adjusting screw counterclockwise to elevate the muzzle of the machinegun; turn the elevating screw clockwise to depress the muzzle of the machinegun.

(2) To adjust the machinegun for traverse—

    (*a*) Loosen the lower socket-head cap screw.

    (*b*) Turn the traverse adjusting screw clockwise to move the machinegun muzzle to the right; turn the traverse adjusting screw counterclockwise to move the machinegun muzzle to the left.

(3) Sight through the bore of the machinegun and aline the axis of the bore on the boresight point by means of the elevating and traverse adjusting screws.

(4) Tighten the socket-head cap screws and reassemble the gun.

h. Removal of Coaxial Machinegun.

(1) Clear the machinegun.

(2) Remove the front and rear locking pins and lift out the machinegun.

13. Cupola Mount, M30

a. General. This mount is designed for use by the tank commander. A caliber .50 machinegun is mounted in the right front portion of the cupola (fig. 1) for use against ground and air targets. The gun can be installed and removed from within the tank (fig. 12). The cupola can be traversed 360° with the traversing handle. It has a maximum elevation of +60° and a maximum depression of −15°. The cupola can be operated from a position within the tank or from a partially exposed position. It can be locked in any position by the friction lock in front of the elevation handle.

b. Mounting Machinegun, Caliber .50, M2E1, Fixed. The gun is mounted in the cupola with the left side of gun down. The mounting is accomplished as follows:

(1) Attach the ejection chute and the link chute to the gun.

(2) Elevate the mount to a 30° angle, and delink the sight arm to prevent damage to the sight linkage.

(3) Remove the rear mounting pin.

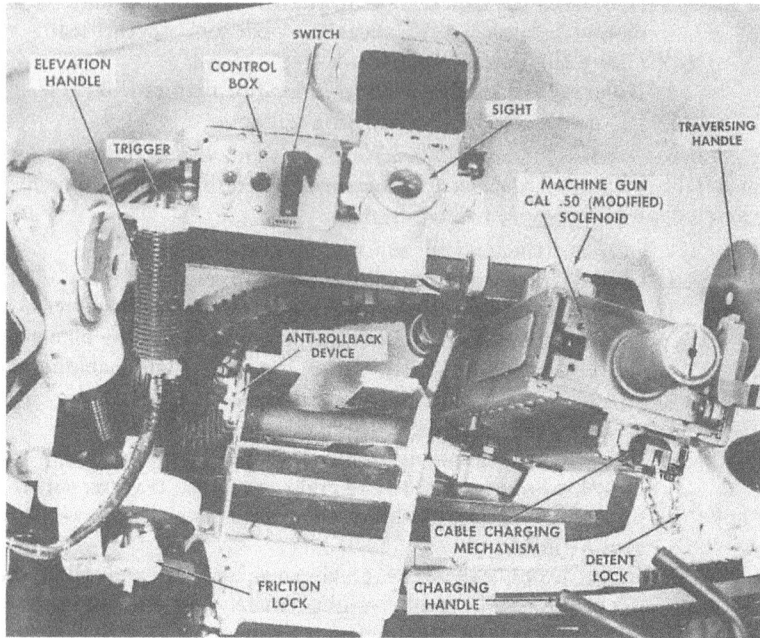

Figure 12. Cupola Mount, M30.

(4) Remove the ammunition box.

(5) Aline the mounting stud, and place the gun in the cupola.

(6) Replace rear mounting pin.

(7) Screw in the barrel.

(8) Aline and attach the feed chute to the feedway.

(9) Connect firing solenoid cable at bell connector.

c. Solenoid Adjustment. The gun is fired electrically with a solenoid; either the G9 or the G12 solenoid may be issued.

(1) *Side plate trigger and G9 solenoid.* The G9 solenoid is mounted on the right side of the receiver. Adjust the solenoid as follows: With headspace properly adjusted, weapon cocked, and the "no fire" gage in place, loosen the right rear screw on the solenoid bracket and turn the knurled adjusting knob clockwise to retard the solenoid plunger until the firing pin cannot be released. Then place the "fire" gage in position and turn the adjusting knob counterclockwise until the sear releases the firing pin. Then turn the adjusting knob two more notches counterclockwise. Tighten the rear right-hand screw on the solenoid. Check final adjustment.

(2) *Back plate (G12) solenoid.* The G12 solenoid is mounted on the buffer tube of the back plate and is adjusted as follows:

26

Adjust headspace and timing according to FM 23–65. Loosen the setscrew on the energizer, and turn the energizer clockwise all the way. Cock the weapon, insert the "fire" gage in proper position, and turn the energizer counterclockwise until the firing pin is released. Recock the weapon and continue turning the energizer counterclockwise, counting each notch, until the firing pin will not release on the "fire" gage. Then turn the energizer clockwise one half the number of notches counted since the firing pin was released, and the solenoid is adjusted. Recheck the adjustment.

d. *Fire-Control Equipment.*

(1) *Periscope sight.* Fire control from within the cupola is provided by a tilting-head monocular periscope sight of 1.27 power with a 30° field of view. The sight reticle is the inverted-T type with three concentric rings and two range dots (fig. 13). The inverted-T is used for boresighting and

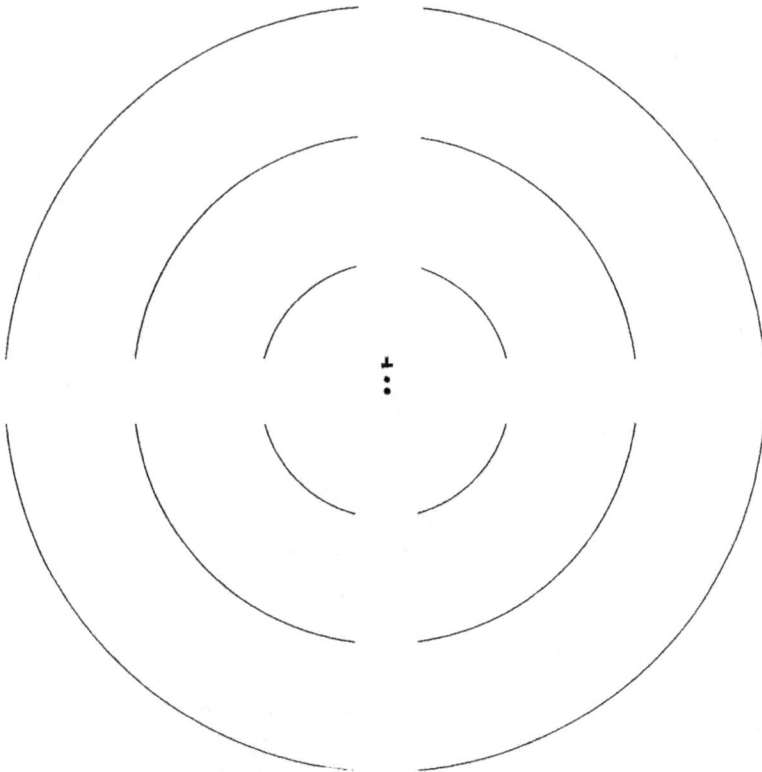

Figure 13. Reticle for periscope, Cupola Mount, M30.

ranges up to 500 yards. The first range dot under the inverted-**T** is for ranges from 500 to 800 yards; the second range dot is for ranges from 800 to 1,000 yards. For targets at ranges greater than 1,000 yards, the tracer stream is used to adjust the fire. The concentric rings are used in engaging aerial targets, each ring being equivalent to 100 miles per hour lead.

(2) *Target-designating device.* An interlock is provided to enable the tank commander to quickly aline the cupola-mounted machinegun with the tank gun. When the cupola is locked by the interlock, the tank commander may use the periscope as an auxiliary target-designating device for the 90-mm gun. Adjustment is made as follows:

(*a*) Engage the interlock by releasing its control handle and swinging the cupola back and forth in the general area of its engagement hole until the lock seats itself.

(*b*) Loosen the three mounting bolts and rotate the cupola until the vertical line of the sight is alined on a given target upon which the main armament has been laid.

(*c*) Tighten the three mounting bolts.

e. Boresighting. Boresighting is accomplished using the inverted-**T** at a range of 500 yards. The aiming point should have clearly defined horizontal and vertical lines. The tank must be placed on level ground to insure accuracy.

(1) *Azimuth adjustment.*

(*a*) Remove the back plate and bolt group from the receiver, and lay the gun on the aiming point by sighting through the barrel.

(*b*) Remove the combination tool from its carrying position in the sight; and with the screwdriver end of this tool, turn the azimuth boresight adjusting screw located on the left side plate of the sight base, to aline the boresight index (inverted-**T**) with the aiming point.

(2) *Elevation adjustment.*

(*a*) Using the same tool, adjust the elevation boresight adjusting screw on the sight drive cam until the inverted-**T** is alined with the aiming point.

(*b*) Make a check of the final adjustment, and assemble the machinegun.

f. Loading. Place the ammunition box in the ammunition box bracket. Secure the box to the bracket with the two clamps at the sides of the box. Pull the ammunition belt through the ammunition feed chute, double-loop end of belt first, and engage the antirollback device. With the cover open, slide the ammunition belt through the feed chute to the feedway of the machinegun by pushing with the fingers at the antirollback device. Close the cover and charge the gun.

g. Firing. Turn the master switch to the ON position. Press the firing trigger on the elevating handle to fire the gun.

h. Locking. The cupola can be locked in any position by a friction type lock located on the front of the cupola directly in front of the elevating handwheel.

i. Dismounting Machinegun. Dismounting is accomplished by performing steps of installation in the reverse order as described in *b* above.

14. Caliber .50 Machinegun Mount, Stock No. 7364875

a. General. This mount is designed for installation on the Tank, 90-mm Gun, M48. The Machinegun, Caliber .50, M2, HB, Flexible, is mounted to the front center of the cupola and is entirely exposed (fig. 2). The gun can be installed and removed only from a partially or fully exposed position. The cupola can be traversed 360°, has a maximum elevation of +60° and a maximum depression of −20°, and can be operated from a position completely within the turret or from a partially exposed position.

b. Mounting of Machinegun, Caliber .50, M2, HB, Flexible. The machinegun has been modified for mounting in the cupola and is installed in the mount in the same manner as in the coaxial type mount, with front and rear mounting pins.

c. Timing, Side Firing Trigger. Timing adjustment for the side firing trigger is accomplished as follows:

  (1) Adjust the headspace (FM 23–65).
  (2) Cock the gun, and loosen the screws holding the bracket to the side of the receiver. Place the "no fire" gage in position, then move the bracket toward the front of the gun until the firing pin will not release (late timing).
  (3) With the "fire" gage in proper position, move the bracket toward the rear of the gun until the firing pin releases sharply.
  (4) To check the timing, tighten the bracket screws and recock the weapon. If the firing pin cannot be released on "no fire" gage, but will release on "fire" gage, the timing is correct.

d. Timing, Trigger Bar Adjusting Nut. This is accomplished as prescribed in FM 23–65.

e. Fire Control. Fire control from within the turret is provided by using the M17 periscope and observing the tracer ammunition. Fire control from the partially exposed position is provided by use of the machinegun sights and observation of tracer ammunition. Manually operated traversing and elevating handwheels are used for manipulation. Firing is accomplished manually from the button-up position through linkage built into the elevation handwheel, and from the partially exposed position by the trigger on the back plate of the machinegun.

f. Use of a Tilting Mechanism on Mount, Stock No. 7364875.

(1) When it is necessary to reload the machinegun from an unexposed position, the commander tilts the machinegun from the firing position into the loader's hatch. The commander first positions the cupola by alining the indicator with the indicator mark in the right rear of the cupola. Using the detachable crank, he unlocks the mount from the cupola by turning the locking shaft located in the right front of the cupola in front of the elevation handwheel.

(2) Using the same crank, he turns the tilting shaft located in the left front of the cupola in front of the traversing handwheel. This tilts the machinegun and mount into position over the loader's hatch, making the left side of the gun and ammunition box accessible to the loader for reloading.

(3) After reloading the gun, it is returned to the firing position by reversing the sequence in (1) and (2) above.

g. Locking. The cupola can be locked in any position by an adjustable friction lock located on the traversing handwheel gear housing. It also can be locked in any position by a positive lock located on the right rear of the cupola.

h. Loading and Charging. Loading the machinegun is accomplished either from the exposed position or by using the tilting feature built into the cupola. Charging the machinegun is accomplished by the commander by use of a cable charger mounted on the left side of the machinegun. A standard caliber .50 ammunition box containing 105 rounds is mounted in the ammunition holder bracket on the left side of the machinegun.

i. Dismounting of Machinegun. Dismounting is accomplished, as in the coaxial type gun, by removing the front and rear mounting pins and disconnecting the firing linkage.

Section IV. TURRET AND ARMAMENT CONTROLS AND EQUIPMENT

15. General

This section describes, locates, illustrates, and furnishes information pertaining to the various turret controls necessary for the proper operation of the turret on the M48 tank.

16. Master Relay Switch

The master relay switch, located on the driver's instrument panel, is the control switch for all the electrical power in the tank. This switch must be in the ON position before the turret can be operated in power, before the guns can be fired electrically, or before the radio interphone system will function.

17. Turret Motor Switch

The turret motor switch is to the left of the power control box (fig. 14). This switch must be in the ON position before the turret can be operated in power. When the switch is on, the indicator light above it will be on.

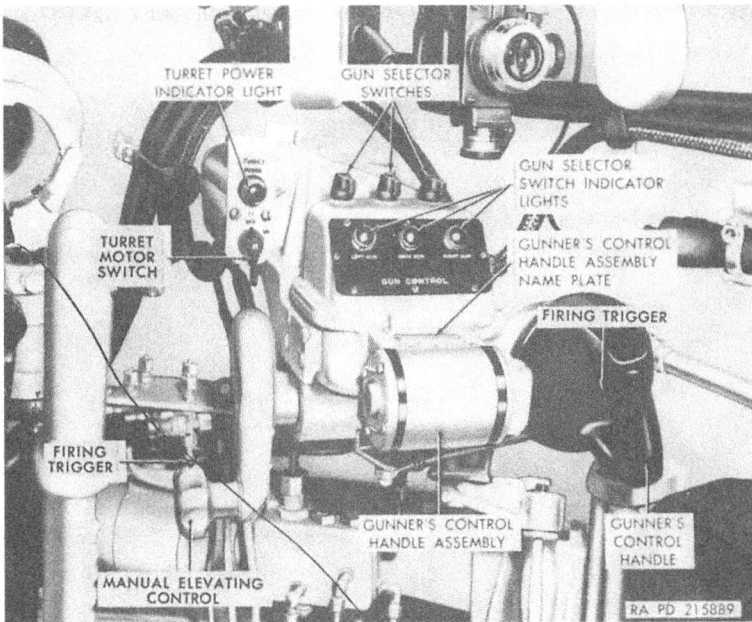

Figure 14. Gunner's power control.

18. Gunner's Power Control

a. General. The gunner's power control handle (fig. 14) is located directly in front of the gunner's position. It provides an integrated system of power control of the turret and tank gun and enables the gunner to traverse the turret and elevate or depress the gun by moving the one control handle.

b. Gunner's Power Traverse Control. The turret can be traversed 360° in either direction as follows: to traverse right, rotate the control handle clockwise; to traverse left, rotate the control handle counterclockwise. The amount the handle is rotated from neutral determines the speed of turret traverse. The maximum speed of turret traverse is four complete revolutions per minute.

c. Gunner's Power Elevation Control. The gun can be elevated to 20° above horizontal and depressed to 9° below horizontal. To elevate, push the bottom of the control handle away from the gunner; to depress, pull the bottom of the control handle toward the gunner. The amount the handle is moved from the neutral position determines the speed of elevation or depression.

19. Tank Commander's Power Control

a. General. The tank commander has the same power control of the turret and gun as does the gunner (fig. 15). The control handle is located on the traversing gear box to the tank commander's right front.

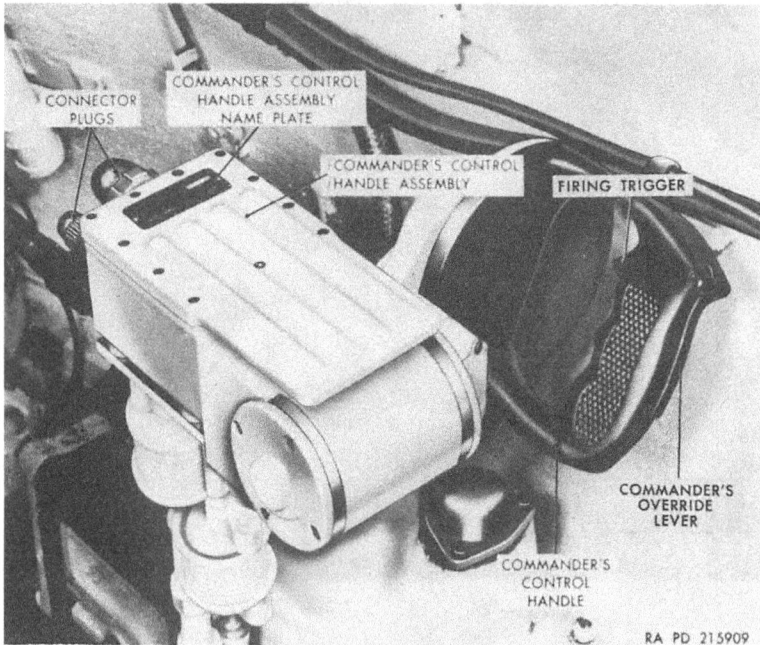

Figure 15. Tank commander's power control.

b. Commander's power Traverse and Elevation Control. The commander obtains power control of the turret by pressing the override lever on his control handle. This action gives him complete control of the gun and turret. To elevate the gun, press the override lever and push the bottom of the commander's control handle away from the commander's position. To depress the gun, press the override lever and pull the bottom of the control handle toward the commander's position. To traverse the turret, press the override lever and rotate the power control handle in the desired direction of traverse.

Caution: Do not press override lever unless handle is centered.
The amount the handle is moved from neutral determines the speed of traverse, elevation, or depression. The gunner does not have power

control of the gun or turret until the commander releases his override lever.

Note. The gun can be elevated or depressed while the turret is being traversed.

20. Gunner's Manual Controls

a. Manual Traverse Control Handle. The gunner's manual traverse control handle is located above and to the right of the gunner's seat. To traverse the turret manually, grasp the manual traverse control handle, squeeze the release lever on the handle, and rotate the handle in the desired direction of traverse. The speed of traverse is regulated by the rate of handle rotation. A no-back mechanism automatically holds the turret in position and prevents turret drift when the vehicle is not in a horizontal position. The gunner can traverse the turret manually when the turret motor is on.

b. Manual Elevation Control Handle. The gunner's manual elevation control handle is located to the left of the gunner's power control handle (fig. 14). It is connected to a hydraulic pump which, by directing a flow of oil to a piston and cylinder assembly, causes the gun to be elevated or depressed. To elevate the gun, rotate the handle clockwise. To depress the gun, rotate the handle counterclockwise. The gun can be elevated and depressed manually while the turret motor is on.

c. Accumulator Hand Pump and Handle. The accumulator hand pump provides an *emergency* means of pressurizing the manual elevation system. The pump is located to the left of the gunner and is operated by means of an accumulator pump handle. To charge the system, move the handle up and down until the gun responds quickly to movements of the manual elevation control handle.

Note. Pressurizing the manual elevation system can be accomplished quickly and effortlessly by turning on the turret motor momentarily.

21. Gun Control Box

The gun control box (fig. 14) is directly in front of the gunner's position. On it are mounted switches and indicator lights that control the electrical firing circuits.

a. The 90-mm gun switch is in the center of the power control box. This switch controls the electrical firing circuit for the main armament and must be in the ON position before the 90-mm gun can be fired electrically. When the 90-mm gun switch is on, the indicator light below it will be on.

b. The coaxial machinegun switch is on the left side of the top of the power control box and controls the electrical firing circuit for the coaxial machine gun. It must be in the ON position before the coaxial machinegun can be fired electrically. When the switch is on, the indicator light below it will be on.

364182 O—55——3

22. Firing Controls

a. Gunner's Firing Controls. The gunner is provided with firing triggers, a hand firing lever, and the gun switches to control the firing of the 90-mm gun and the coaxial machinegun. Firing triggers which permit electrical firing by the gunner are located as follows: one on the front of the manual elevation control handle, and one on the front of the gunner's power control handle. The hand firing lever is located on the upper right side of the gun cradle, and is easily accessible from the gunner's position. This lever permits manual firing on the 90-mm gun.

b. Commander's Firing Controls. The tank commander may fire the 90-mm gun or the coaxial machinegun by pressing the override lever and using the firing trigger on the front of the tank commander's power control handle. The firing trigger for the tank commander provides electrical firing only. It will not function unless the necessary switches are on and the commander has taken power control of the turret by pressing the override lever on his power control handle.

c. Safety. The 90-mm safety is on the top left side of the gun mount (fig. 16). When the safety is to the rear, the gun can be fired. When the safety is forward, the firing plunger is blocked, and the gun cannot be fired.

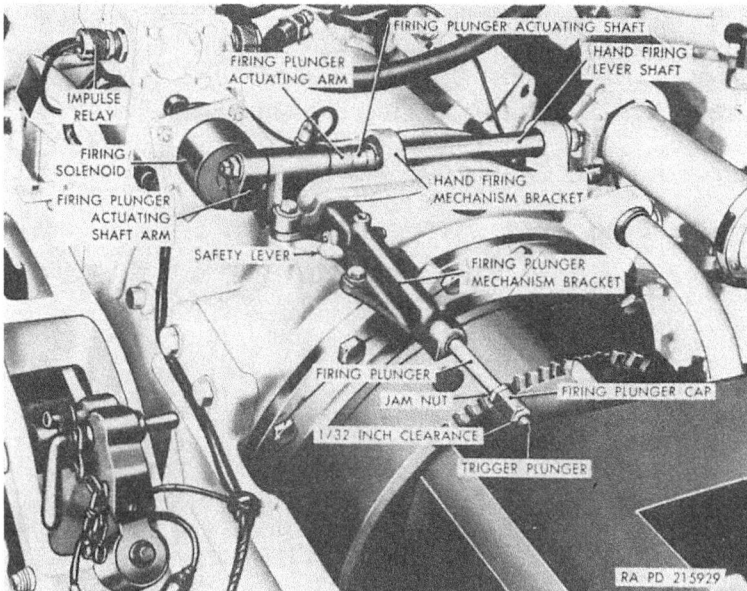

Figure 16. Manual safety and firing linkage.

23. Locks

a. Turret Lock.

(1) The turret lock, located to the right of the gunner's seat, holds the turret stationary by means of a gear segment which engages the turret ring teeth. The turret lock should be in the locked position while the vehicle is in motion, unless the turret is to be traversed.

(2) To unlock the turret lock, turn the turret lock handle counterclockwise until the gear segment disengages from the turret ring gear.

(3) To lock the turret, turn the handle clockwise until the turret will not traverse.

Caution: Do not use power traverse to check the turret lock.

b. Gun Traveling Lock.

(1) The gun traveling lock, located on the rear deck of the tank, is used during periods of nontactical operations to keep the gun in the locked position, thereby avoiding excessive wear of the traversing and elevating mechanism while the tank is moving.

(2) To open the lock, lift the lever from the top of the cap and unscrew the lever bolt from its bracket by turning the lever counterclockwise. Swing the cap and lever back. Lay the traveling lock flat on the top deck. Before the traveling lock can be raised to the vertical position again, the stop pawl must be released. To close the lock, swing the cap over the gun tube and tighten the lever by turning it clockwise. Rotate the lever into position over the cap and push the lever down.

24. Putting Turret Into Power Operation

The following steps are performed in putting the turret into power operation. See paragraph 66 for individual crew duties. Remember the word ACUTE as the key.

A. *Alert crew*_____ Insure that the crew is in safe position. Check tank and area for obstructions.

C. *Check oil*_____ Oil should be at FULL mark on bayonet gage.

U. *Unlock turret*_____ Traverse manually to make sure turret is unlocked. Return manual traverse control handle to latched position.

T. *Turn on power*_____ Power control handles should be in neutral position.

E. *Elevate and traverse*_____ Operate in power to make sure controls are functioning properly.

Note. The main or auxiliary engine should be running to charge the tank batteries while the turret is in power operation.

Section V. DIRECT-FIRE SIGHTS, VISION DEVICES, AND AUXILIARY FIRE-CONTROL EQUIPMENT

25. General

The Tank, 90-mm Gun, M48, is equipped with primary and secondary means for direct fire, necessary instruments for indirect laying of the tank gun, and the vision devices required by the crew for operation of the tank with all hatches closed. This section contains information concerning location, adjustment, and use of these various pieces of equipment.

26. Direct-Fire Sights

a. General. Direct-fire sighting systems permit precise laying of the tank gun on a target which can be seen from the tank. Each direct-fire sight contains a gun-laying reticle. The sight is mounted in such a manner that the reticle moves in elevation and deflection to correspond to movement of the gun. The reticle may also be moved independently of the gun by use of the boresight knobs. Reticles may be illuminated for use during darkness.

b. Primary Direct-Fire Sighting System. The primary direct-fire sighting system in the M48 tank consists of Periscope, M20 (M20A1 or M20A2), in Periscope Mount, T184; Ballistic Drive, T24E2; Range Finder, T46E1; and Computer, T31.

o. Secondary Direct-Fire System. The secondary direct-fire system is the Telescope, T156E1, in Telescope Mount, T191.

27. Periscope, M20, M20A1, or M20A2

a. The Periscopes, M20, M20A1, and M20A2 (fig. 17) are similar and interchangeable. For the purpose of this manual, they are treated as being the same and are referred to as M20 only. The M20 periscope, a single-eyepiece instrument, has two built-in optical systems: a one-power system for wide-angle, close-in observation, and a six-power system for sighting. The periscope is composed of two main parts: a replaceable head assembly which is attached to the top of the periscope mount, and a body assembly which is secured to the bottom of the mount. A coupling assembly with an elevation lever connects the movable mirror in the periscope head to a coupling on the shaft of the ballistic drive. The shaft of the ballistic drive also is connected to the gun through linkage. When the gun is elevated or depressed, the mirror in the periscope head moves, causing the gunner's line of sight through the periscope to be elevated or depressed the same amount as is the gun. The line of sight of the periscope may be moved independently of the gun by action of the computer.

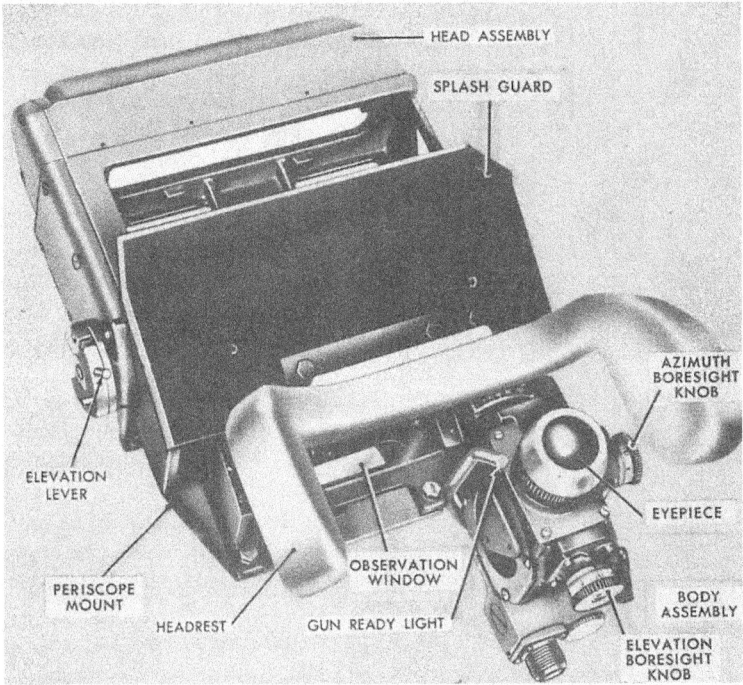

Figure 17. Periscope, M20.

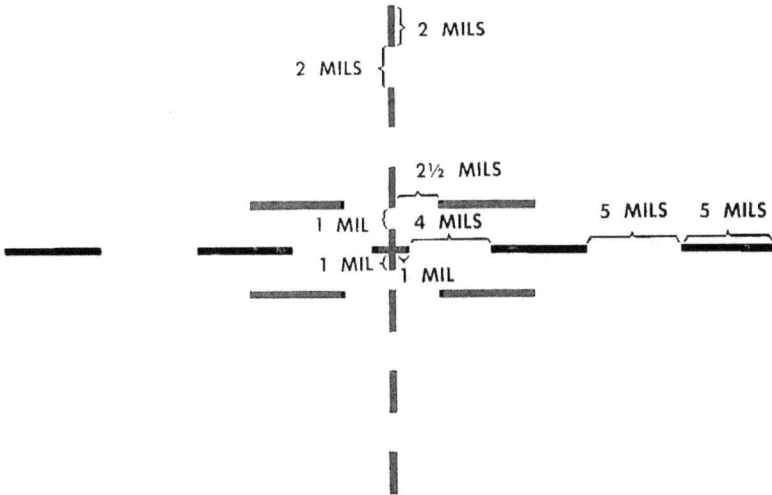

Figure 18. Gun-laying reticle.

37

b. A reticle pattern (fig.18) is contained in the periscope six-power system. It is composed of horizontal and vertical lines, and spaces, for measuring mil angles as indicated in figure 18. The aiming cross, in the center of the reticle, is used for sight adjustment and for firing the initial round at stationary targets. The vertical lines are called "range" lines, and the horizontal lines are called "lead" lines. The one-power system has no reticle pattern.

c. The elevation lever (fig. 17) on the coupling allows the line of sight through the periscope to be elevated independently as much as 22 degrees from zero elevation. This increases the amount of terrain visible through the periscope without elevating the gun. The lever must be held in the depressed position when installing the periscope head.

d. A diopter adjustment is provided on the eyepiece of the six-power system for focusing the periscope to the eye of the user. The diopter scale is graduated from −3 to +3 diopters. This facilitates resetting the adjustment once the proper setting is known.

e. Two boresight knobs are provided on the periscope body to move the reticle during sighting. Each knob has a slip scale graduated from .5 to 5.5 in mils for recording sight adjustments. Locking levers are provided to hold the adjustments.

f. The reticle of the M20 periscope is illuminated by means of an instrument light. A dovetail slot over the eyepiece of the six-power system receives the lamp head of the instrument light.

g. A splash guard is installed on the back of each mount for protection of the crew against shell fragments and small-arms projectiles in the event the periscope head is hit. Two retaining fasteners hold the guard on the mount.

28. Ballistic Drive, T24E2

a. The ballistic drive (fig. 19) consists of a ballistic unit and the necessary linkage to connect it to the 90-mm gun and to the periscope, computer, and range finder.

b. Changes in elevation of the gun cause a corresponding change in the line of sight in the periscope and the range finder. This change is transmitted through the elevation linkage and the elevation shaft to the ballistic unit, and then to the periscope and the range finder. The superelevation angle (fig. 20) is transmitted from the computer through the ballistic drive input shaft to the ballistic unit. This movement causes the ballistic drive input shaft to rotate independently of the elevation shaft. This action changes the superelevation angle by moving the mirror in the head of the M20 periscope, thus deflecting the line of sight to a new angle without moving the gun. To bring the line of sight back to the aiming point, with the new superelevation angle, it is necessary to use the gun-elevating controls.

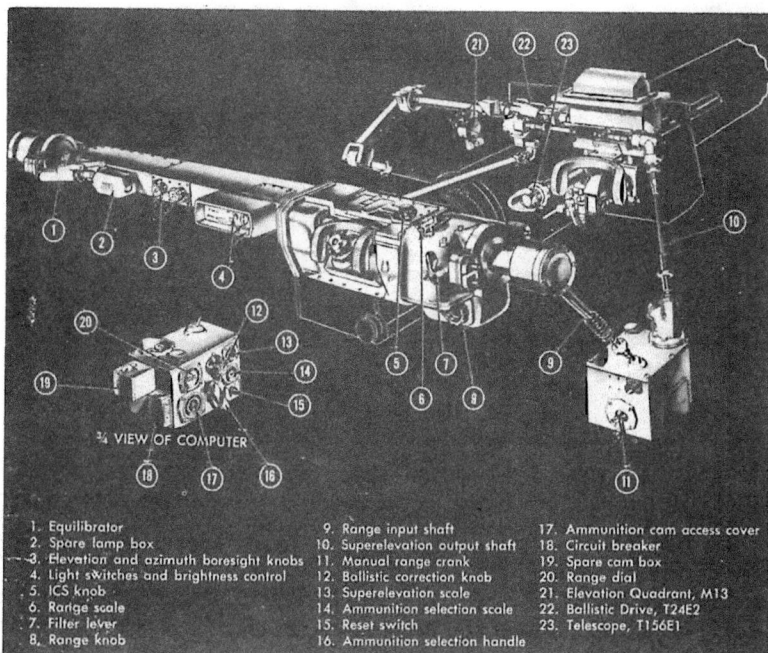

Figure 19. *Fire control instruments and Ballistic Drive, T24E2.*

1. Equilibrator
2. Spare lamp box
3. Elevation and azimuth boresight knobs
4. Light switches and brightness control
5. ICS knob
6. Range scale
7. Filter lever
8. Range knob
9. Range input shaft
10. Superelevation output shaft
11. Manual range crank
12. Ballistic correction knob
13. Superelevation scale
14. Ammunition selection scale
15. Reset switch
16. Ammunition selection handle
17. Ammunition cam access cover
18. Circuit breaker
19. Spare cam box
20. Range dial
21. Elevation Quadrant, M13
22. Ballistic Drive, T24E2
23. Telescope, T156E1

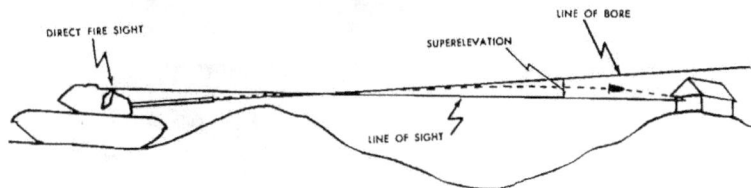

Figure 20. *Superelevation.*

29. Range Finder, T46E1

a. General. The Range Finder, T46E1 (figs. 19, 21, and 22), is secured to the turret roof in rear of the ballistic drive. It is a stereoscopic range finder, operated by the commander as part of the primary sighting and fire-control system. It mechanically transmits range data to the Ballistic Computer, T31.

b. Data.

Operating temperature range_____ —65° F. to +150° F.
Operating voltage_____ 24 volts dc.
Range_____ Graduated from 500 to 4,800 yards.
Magnification_____ 10x.
Field of view_____ 4° (71 mils).
Base length_____ 79 inches.

39

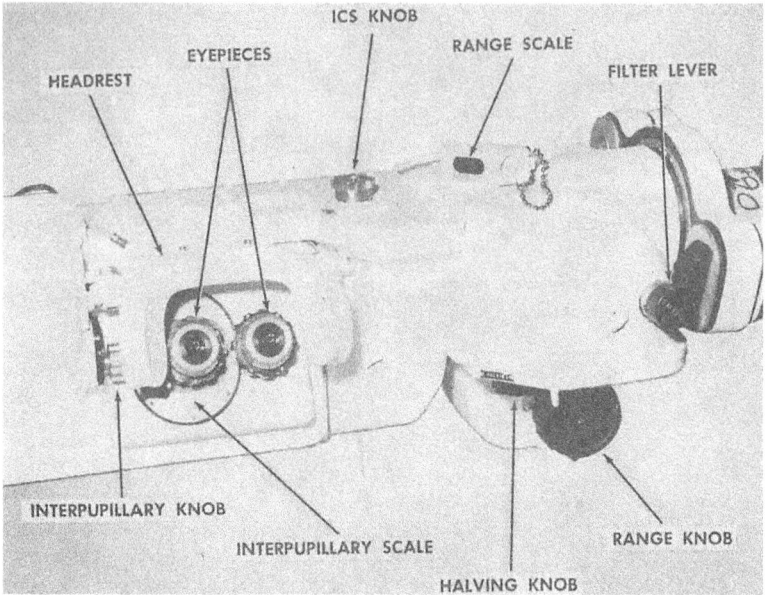

Figure 21. Range Finder, T46E1.

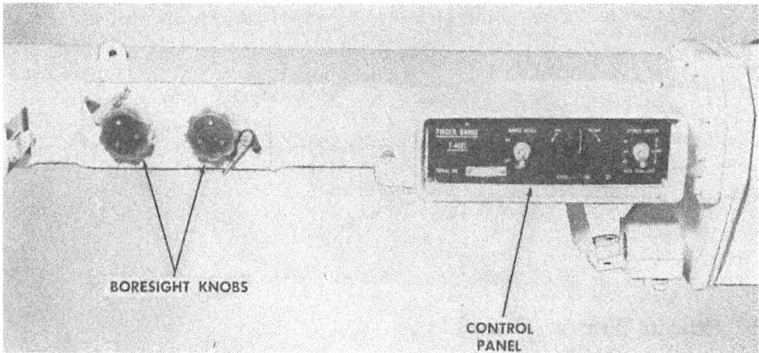

Figure 22. Range Finder, T46E1.

c. Reticle Patterns.

   (1) *Gun-laying reticle.* The gun-laying reticle (fig. 23) appears in the center of the field of view. It is similar to the reticle in the M20 periscope, except that there are only two range lines.

   (2) *Stereoscopic pattern (ranging reticle)* (fig. 24). When illuminated, a ranging reticle is viewed through each eyepiece. Each reticle is formed by five vertical lines placed to form a ||¦|| (V) pattern. This pattern is used to gain stereo-

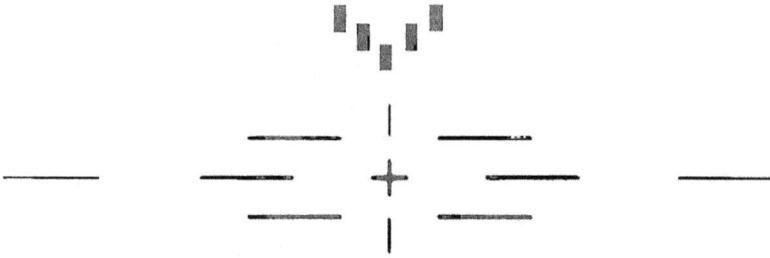

Figure 23. Reticles, Range Finder, T46E1.

scopic contact with the target. With the range finder in proper adjustment, the ranging reticles will fuse, forming a pattern which appears to have depth. The lowest bar appears to be at a greater range than the two center bars; the two upper bars appear to be at a closer range than the two center bars. As the range knob is rotated, the ranging reticle will appear to move in depth. When ranging on a target, the range knob is rotated until the lowest vertical bar of the ranging reticle appears to be at the same range as the target.

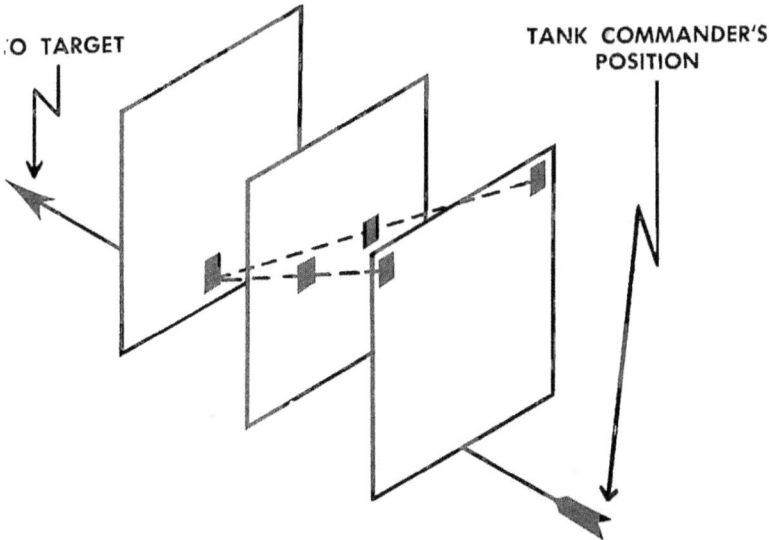

Figure 24. Ranging reticle (shows depth of reticle).

(3) *Auxiliary gun-laying reticle.* The auxiliary gun-laying reticle must be illuminated in order to appear in the field of view. It is identical to the primary gun-laying reticle.

41

d. Controls (figs. 21 and 22).

    (1) *Diopter scales.* Each eyepiece has a diopter scale which is used to focus the eyepieces to the tank commander's eyes. The scales are graduated from -4 to $+4$ diopters. Adjustment or focusing is accomplished by rotating the eyepiece.

    (2) *Interpupillary knob.* The interpupillary knob, located to the left of the eyepieces, enables the observer to adjust the distance between the eyepieces to correspond to the distance between his eyes. An interpupillary scale, graduated from 58 to 72 millimeters, is provided so that the observer may record his setting, to facilitate adjusting the eyepieces once his setting is known.

    (3) *Halving knob.* The halving knob is located above the range knob. This knob permits vertical adjustment of the right ranging reticle. The halving adjustment is made when the left and right ranging reticles appear at different elevations within the field of view (fig. 25).

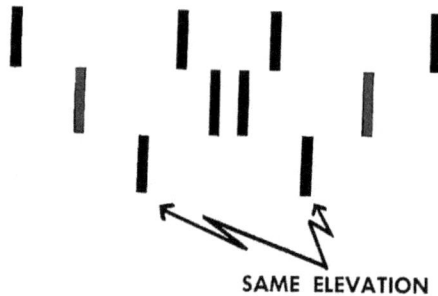

SAME ELEVATION

Figure 25. Ranging recticle (proper halving adjustment) (improper halving adjustment).

(4) *Internal correction system knob.* The internal correction system (ICS) knob is located on top of the range finder to to the right of the eyepiece assembly. The knob is provided with a scale which is graduated from 0 to 50 in units of error. It permits the adjustment of the internal mechanism to the eyes of the tank commander. The setting will vary for each instrument and for each tank commander. See paragraph 35c(10) for procedure in determining the individual ICS setting.

(5) *Filter lever.* The filter lever, located to the right of the eyepieces, controls filters for the left and right optical systems of the range finder. When the lever is positioned to the right, the filters are removed from the field of view; when the lever is moved to the left, the filters are introduced into the optical system.

(6) *Boresight knobs.* The T46E1 range finder has two sets of boresight knobs, which are used to adjust the gun-laying reticles when boresighting. One set of boresight knobs is located on the left front of the main housing and controls the gun-laying reticle movement seen through the left eyepiece. The second set of boresight knobs is located on the right rear of the main housing and controls the auxiliary gun-laying reticle movement seen through the right eyepiece. In each set of knobs, the knob to the operator's left controls the reticle in azimuth, and the knob to the operator's right controls the reticle in elevation. An adjustable scale is around each boresight knob and can be rotated independently of the knob. A locking lever is provided on each knob to insure that the desired setting will not be changed by vibration during tank operation.

(7) *Light switches and rheostat.* The stereo switch is a three-position toggle switch located on the right side of the control panel. The three positions are marked from the top as follows: ON-STEREO SWITCH, OFF, and ON-AUX. GUNSIGHT. This switch controls the illumination of the ranging reticles and the auxiliary gun-laying reticle. Brightness of the illuminated reticles is adjusted by rotating the rheostat knob located to the left of the stereo switch. The range scale switch is located to the left of the rheostat.

(8) *Range knob and scale.* The range knob is located to the right of and slightly lower than the eyepieces. It is used to control the apparent movement of the ranging reticle or to index a given range on the range scale. The range scale is located on top of the range finder to the right of the ICS knob. It is graduated from 500 to 4,800 yards and moves past its fixed index when the range knob is rotated.

(9) *Lamps*. Provision is made to illuminate the various reticles and scales. The lamp for the right ranging reticle is located in a receptacle on the left front of the range finder. A similar receptacle for the left ranging reticle is located to the left of the auxiliary gun-laying reticle boresight knobs. The lamp for the auxiliary gun-laying reticle is located in a receptacle to the left of the lamp for the left ranging reticle. The range scale lamp, in a receptacle on the range scale housing, is controlled by a switch on the left side of the control panel. These lamps may be replaced when necessary by unscrewing the knurled plug.

(10) *Spare lamps*. Spare lamps are carried in the space lamp box located on the left rear of the range finder. There is space for four reticle lamps (the ranging reticle lamps and auxiliary gun-laying reticle lamps are identical) and two range scale lamps.

e. Operation. The following steps are necessary to put the range finder into operation. Due to the functional relationship of the various controls, it is recommended that these steps be performed in the order listed.

(1) *Diopter adjustment*. The tank commander looks through the instrument, using both eyes. He adjusts each eyepiece individually, with the other blacked out, until the image in each eyepiece is seen with maximum sharpness. The correct diopter setting will be indicated on the diopter scales. These settings should be noted and memorized for future use.

(2) *Interpupillary adjustment*. The tank commander first determines his interpupillary distance with a binocular. Noting the setting on his binocular, he indexes this setting on the interpupillary knob, and memorizes it for future use.

(3) *ICS adjustment*. The tank commander indexes his individual ICS setting on the ICS scale. In the event the tank commander has not determined his ICS setting, he uses a setting of 25.

(4) *Lights and filter adjustment*. The tank commander turns the stereo switch to the ON-STEREO SWITCH position. He introduces the filter if desired. Next, he adjusts the brightness of the reticles with the rheostat. The intensity of light should be reduced to the minimum which will provide well-defined reticles.

(5) *Halving adjustment*. The halving adjustment is accomplished by either of the following methods. In the first, or binocular, method, the tank commander sets the range scale at maximum range. Without looking through the eyepieces, he depresses the gun to minimum elevation. The

ranging reticle then will appear as two separate patterns (fig. 25). With both eyes open, he moves the right reticle with the halving knob until it appears to be at the same elevation as the left. In the second method, which is less accurate, he closes his right eye and, by use of the gun controls, lays the bottom of the left ranging reticle on a target at about 1,500 yards. Closing his left eye, he opens his right eye, and, with the halving knob, lays the bottom of the right ranging reticle on the same target. He then closes and opens his eyes alternately to recheck the adjustment.

30. Computer, T31

a. General. The Computer, T31 (figs. 19 and 26), is an electromechanical device designed to compute superelevation angles (fig. 20) for the 90-mm gun. When operated electrically, the computer receives range data from the range finder. This range data then is applied to the ammunition data and ballistic corrections which previously have been indexed manually into the computer by the gunner. The product is the superelevation angle, which is transmitted through the ballistic drive to the M20 periscope. The ballistic drive also transmits the same superelevation angle to the range finder. A superelevation crank is provided for manual introduction of the superelevation angle.

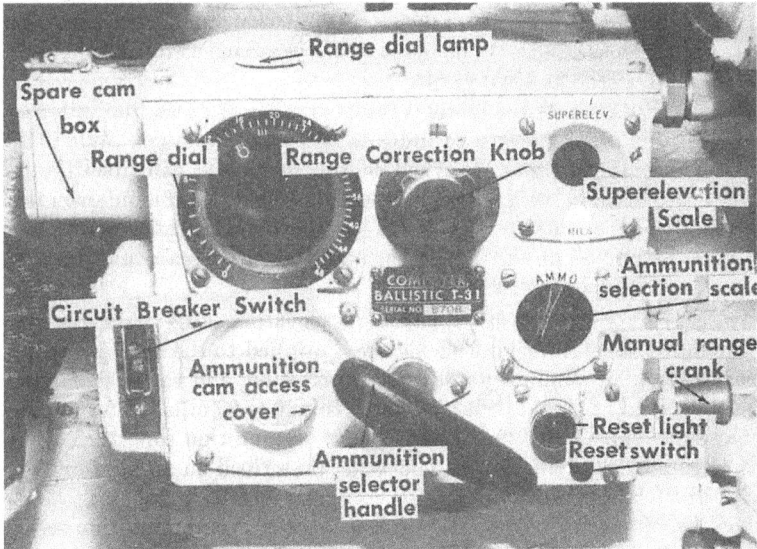

Figure 26. Computer, T31.

b. Switches and Lamps.

    (1) *Stereo switch.* The stereo switch on the control panel of the range finder must be in the ON-STEREO SWITCH or ON-AUX. GUNSIGHT position to supply electricity for automatic operation of the computer. The range dial lamp should light when the stereo switch is in either position.

    (2) *Circuit breaker.* A circuit breaker switch is provided on the left of the computer. If an electrical short occurs in the computer, or the computer is overloaded, the switch is automatically kicked to the OFF position. When this happens, it is necessary to manually return the switch to the ON position. If the switch will not stay on, notify maintenance personnel.

    (3) *Reset button and light.* When the gunner operates the superelevation crank, the computer ceases to function electrically, and a red light next to the reset button will glow. The reset button must be pushed to restore electrical operation.

    (4) *Range dial lamp.* The purpose of the lamp is to illuminate the range dial. A defective lamp may be replaced with a new one from the spare lamp box on the range finder.

c. Range Dial. The computer range dial is located on the face of the computer. The dial is fixed and is graduated from 0 to 4,800 yards. Two movable pointers operate in conjunction with the dial.

    (1) When the range knob of the range finder is rotated, the index on the inner pointer will indicate the same range as that indexed on the range finder (except when range correction has been applied). The index will move from a stop at 500 yards to a stop at 4,800 yards.

    (2) Rotation of the manual range crank will cause the index on the outer pointer to indicate the desired range. When the computer again is turned on, this index will aline itself automatically with, and follow any movement of, the index on the inner pointer. The index on the outer pointer will move from a stop at 0 yards through the entire scale and back to a stop at 0 yards.

d. Superelevation Scale. The superelevation scale indicates the amount of superelevation that has been applied to the M20 periscope for the type of ammunition, ballistic correction, and range that have been indexed. This scale is graduated from 0 to 99 mils. The gunner may set in superelevation for any type of ammunition for which there is no cam by applying the mil elevation prescribed on an aiming data chart or firing table.

e. Ammunition Selector Handle and Scale. An ammunition selector handle is provided so that the gunner may index any one of six types of ammunition. This is necessary since velocity and flight characteristics of different types of ammunition vary considerably. The handle first must be rotated clockwise approximately one-eighth turn and then pushed or pulled until the tab identifying the type of ammunition desired is indexed on the ammunition selector scale.

(1) *Ammunition cams.* Eight different ammunition cams are provided of which only six are installed. The remaining two are stored in the spare cam box. With this arrangement, it is possible for the gunner to have any six of the eight cams installed and ready for use. Each cam has etched on its surface the type of ammunition for which it is graduated. The gunner may change these cams (fig. 27) by following the steps outlined below.

Figure 27. Computer, T31.

(a) From the spare cam box, remove the cam to be installed.

(b) Using the ammunition selector handle, index the tab for the cam that is to be replaced.

(c) Remove the four screws from the ammunition cam access cover, and remove the cover.

(d) Turn the ammunition selector handle to its full clockwise position and continue to hold it in this position while performing steps (e) and (f) below.

Caution: It is important that the handle be held as described in (d) above.

47

(e) Remove the cam that is to be replaced. (It may be necessary to remove a number of cams to gain access to the cam being replaced.)

(f) Insert the new cam, and replace the remaining cams in proper order.

(g) Release the ammunition selector handle, and replace the ammunition cam access cover.

(2) *Ammunition tabs.* When a new ammunition cam is installed, the tab for the new cam must be installed in the ammunition selector scale. This is accomplished as follows:

(a) Make sure that the ammunition selector handle is pulled to the proper position.

(b) Remove the four screws from the ammunition selector scale cover, and remove the cover.

(c) Remove the tab which identifies the old cam, and install the new tab.

(d) Replace the ammunition selector scale cover.

f. Range Correction Knob. Range corrections may be made to compensate for variable conditions which affect gun or ammunition performance—e. g., air density and temperature, powder and tube temperature, tube wear, variation in lots of ammunition, etc. The knob is graduated in percent of range, from −15 to +15. Until such time as appropriate data is available, this knob will be set at zero.

g. Superelevation Crank. The superelevation crank is used by the gunner to introduce manually the superelevation angle to the periscope and range finder. It is used in the event of an electrical failure and when boresighting. The crank is spring-loaded and must be held against the spring tension while being used. It is rotated until the outer range dial pointed matches the inner pointer, or until the desired angle is indexed on the superelevation scale. To return the computer to electrical operation, it is necessary to push the reset button.

31. Telescope, T156E1, and Telescope Mount, T191

a. Telescope, T156E1.

(1) The Telescope, T156E1 (fig. 28), is provided as a secondary direct-fire system to be used in the event the primary direct-fire sighting system becomes inoperable.

(2) The instrument is held coaxially with the gun by the telescope mount, which is attached to the right side of the recoil cylinder. The telescope has an eight-power optical system with a 132-mil, circular field of view. A diopter scale graduated from −3 to +3 diopters is provided on the eyepiece, for focusing the telescope to the gunner's eye. Wing nuts on the eyeshield bracket allow the user to adjust the distance between the eyeshield and eyepiece.

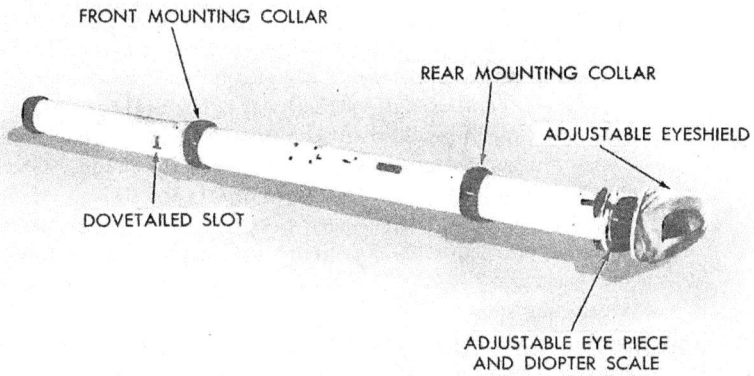

Figure 28. Telescope, T156E1.

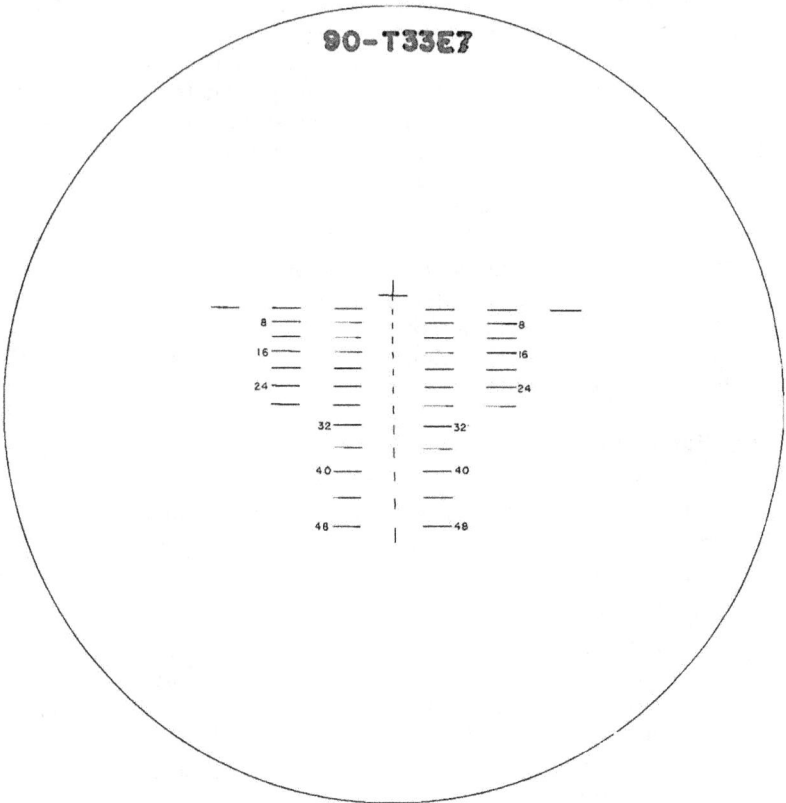

Figure 29. Reticle pattern, Telescope, T156E1.

(3) The reticle (fig. 29) in the telescope is a standard dash-pattern type, graduated from 0 to 5,000 yards for the 90-mm, APT-T33E7 round. The intersection of the lines forming the boresight cross represents zero range and zero deflection. The reticle is composed of vertical and horizontal lines. Each of the verticle lines, known as range lines, and each space between the range lines, represents 200 yards of range. Graduations are numbered at 800-yard intervals. Each of the horizontal lines, and the space between the lines, measures 5 mils or one lead. When firing any type of ammunition other than APT-T33E7, it will be necessary to refer to an aiming data chart.

(4) Illumination of the reticle is possible through use of an instrument light. A dovetailed slot forward of the front mounting collar receives the head of the instrument light.

b. Telescope Mount, T191. The Telescope Mount, T191 (fig. 30), consists of an adjustable bracket which receives the rear mounting collar of the telescope, and an aperture in the gun shield which receives the front mounting collar. To install the telescope, unscrew the wing nut on the adjustable bracket, and lift the upper half of the bracket. Slip the front mounting collar into the gun shield aperture, and position the rear mounting collar in the adjustable bracket, making sure that the locating stud is positioned in its recess. Close the upper half of the bracket and secure with the wing nut. Adjustment of the telescope during boresighting or zeroing is accomplished by means of the elevation and azimuth boresight knobs on the telescope mount. The knobs are held in any set position by locking levers. Scales, graduated from 0 to 4 mils, are provided on the knobs to permit indexing and recording sight adjustments. The scales may be "slipped" by loosening the wing nut and rotating the knob.

32. Boresighting

The purpose of boresighting is to establish a definite relationship between the axis of the gun tube and the lines of sight of the direct-fire sights.

a. Preliminary Steps.

(1) Position the tank as level as possible.

(2) Insert the breech boresight into the chamber, or remove the percussion mechanism and use the firing pin well as a breech boresight. Tape black thread to form a cross over the witness lines on the blast deflector or on the muzzle of the tube.

(3) Perform the following steps to place the range finder into operation: Adjust diopter setting; make interpupillary adjustment; place individual ICS setting on the ICS scale;

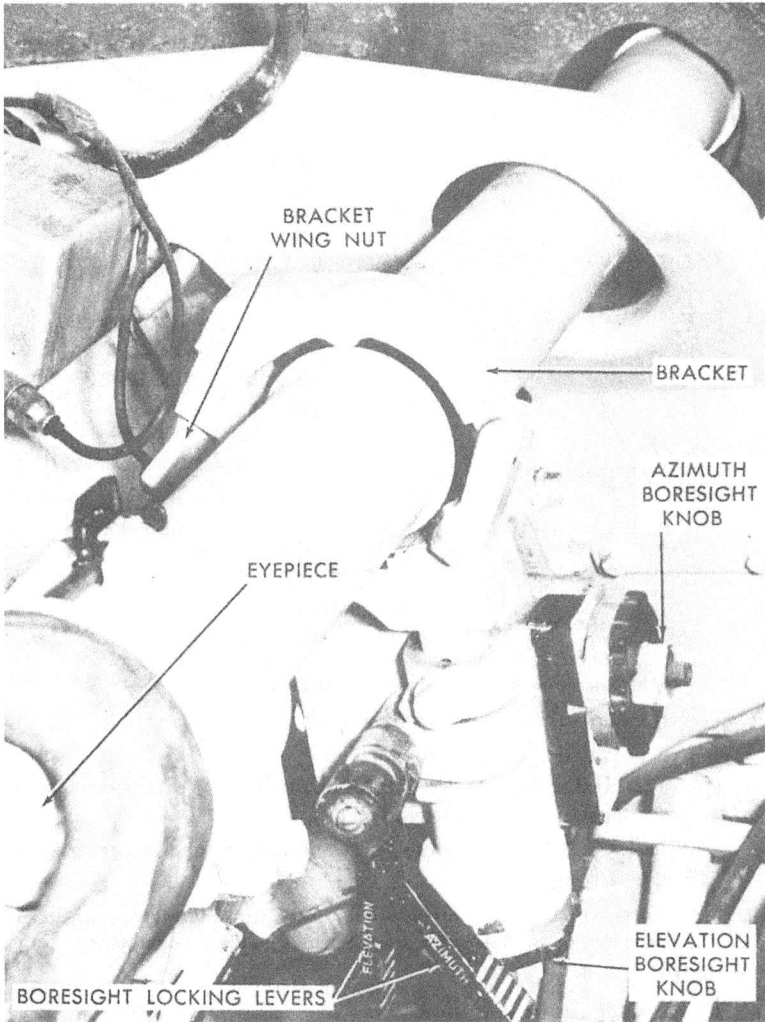

Figure 30. Mount, Telescope, T191.

check reticle lights; position filter, if needed; and adjust halving (par. 29).

(4) Select a target as near 1,500 yards as possible (preferably a target with clearly defined horizontal and vertical lines).

(5) Position the right telescope of a binocular over the firing pin well. Using the firing pin well (breech boresight) as a rear sight and the cross threads on the muzzle as a front sight, aline the axis of the bore on the boresight point by use of the manual traverse and elevation controls (fig. 31).

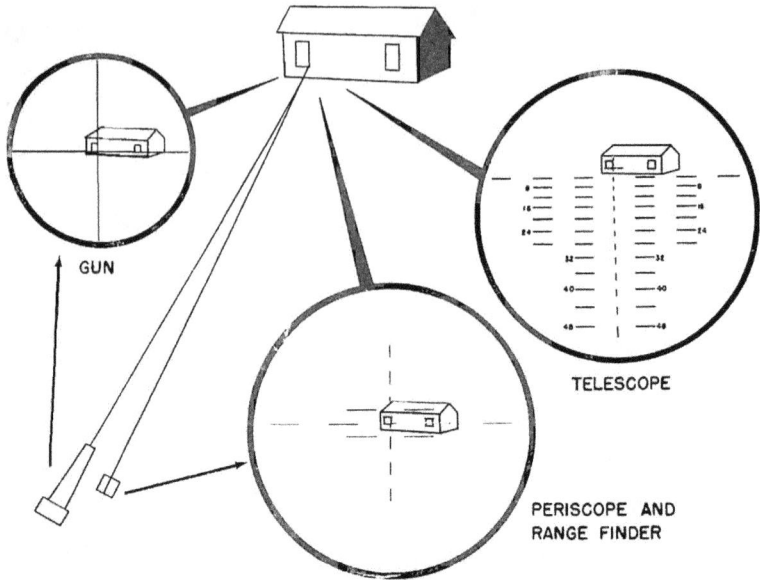

Figure 31. Boresighting.

> *Note:* Make sure that the accumulator pressure is high enough to hold the gun on the boresight during subsequent steps.

(6) Remove all superelevation as follows:

  (*a*) Turn the computer off.

  (*b*) Index zero on the superelevation scale of the computer by use of the superelevation crank.

(7) Index the known range to the target on the range scale of the range finder.

> *Note:* The following steps are normally accomplished at the same time. For better understanding they are listed separately.

b. Periscope, M20.

(1) Unlock the boresight knobs.

(2) Sighting through the periscope, rotate the boresight knobs as necessary to move the aiming cross exactly on the boresight point (fig. 31).

(3) Lock the boresight knobs.

(4) Recheck to insure that the muzzle cross threads and periscope aiming cross are on the boresight point.

(5) Turn the slip scales to read "4" on the elevation boresight knob and "4" on the deflection boresight knob.

c. Range Finder, T46E1.

(1) Unlock the gun-laying reticle boresight knobs.

(2) Rotate the gun-laying reticle boresight knobs to move the timing cross of the gun-laying reticle to the boresight point (fig. 31).

(3) Lock the gun-laying reticle boresight knobs.

(4) Recheck to insure that the muzzle cross threads and the gun-laying aiming cross are on the boresight point.

(5) Turn the slip scales to read "4" on the elevation boresight knob and "4" on the azimuth boresight knob. (Some earlier models have slip scales numbered in reverse, requiring a setting of "2" and "2".)

(6) Adjust the auxiliary gun-laying reticle by turning the stereo switch to the ON–AUX. GUNSIGHT position; and, using the auxiliary gun-laying reticle knobs, proceed as for the gun-laying reticle ((1) through (5) above).

  d. *Telescope, T156E1.*

(1) Unlock the telescope boresight knobs.

(2) Rotate the boresight knobs to move the boresight cross of the telescope reticle to the boresight point (fig. 31).

(3) Lock the telescope boresight knobs.

(4) Recheck to ensure that the muzzle cross threads and the telescope boresight cross are on the boresight point.

(5) Loosen the wing nuts on the boresight knobs and rotate the knobs to index "2". Tighten the wing nuts.

33. Zeroing

  a. *General.* The purpose of zeroing is to cause the projectile of a given type ammunition to hit the point of aim at a given range.

  b. *Steps in Zeroing.*

(1) *Boresighting.* Boresight as outlined in paragraph 32.

(2) *Emergency zero.* The emergency zero is applied to the direct-fire sights before zeroing when a zero has not been previously established. To apply the emergency zero proceed as follows: Unlock the elevation and azimuth boresight knobs. Then, being careful not to move the slip scales, rotate each knob until the settings listed below are indexed.

	Elevation	*Deflection (azimuth)*
Periscope	3	3
Range Finder	3	3
Telescope	Leave at boresight until zeroing is completed; then refer as described in (14) below.	

Lock all boresight knobs.

> *Note.* The telescope normally will be made to coincide with the M20 periscope after the zero for the periscope has been determined. If a zeroing exercise cannot be fired, the range and ammunition data should

be set into the computer, and the gun re-laid to place the aiming cross on the boresight aiming point. The telescope boresight knobs, being unlocked, are rotated to move the appropriate point on the range line in the telescope reticle to the same aiming point.

(3) Select a target with a clearly defined aiming point as near 1,500 yards as possible. (The boresight target will normally be used.)

(4) Turn the computer on.

(5) Index the ammunition on the computer. (Shot ammunition should be used.)

(6) Index zero on the range correction knob.

(7) Index the known range to the target on the range scale of the range finder.

(8) Check to insure that the range pointers of the computers are matched, then turn the computer off.

(9) Using the manual elevation and traverse controls, lay the aiming cross of the periscope on the aiming point of the zeroing target (fig. 32).

(10) Fire three to five rounds (all with the same lot number) to form a shot group on the target. Check the lay of the gun after firing each round; re-lay if necessary (fig. 32).

(11) Without disturbing the lay of the gun, unlock the periscope boresight knobs and move the aiming cross of the gun-laying reticle to the center of the shot group. Relock the periscope boresight knobs (fig. 32).

(12) Unlock the gun-laying reticle boresight knobs on the range finder, and move the aiming cross to the center of the shot group. Lock the boresight knob (fig. 32).

(13) Using the manual elevation and traverse controls, re-lay on the aiming point of the zeroing target. Fire a round to check the accuracy of the settings. (The projectile should hit within 14 inches of the aiming point.)

(14) Re-lay the aiming cross of the periscope on the aiming point. Unlock the boresight knobs of the telescope and the auxiliary gun-laying reticle of the range finder. Rotate the boresight knobs to move the appropriate range point of the telescope reticle and the aiming cross of the range finder auxiliary gun-laying reticle on the aiming point. Lock the boresight knobs.

(15) Record all boresight knob settings in some convenient place in the tank turret and, in pencil, in the gun book. These settings are the established zero settings. In subsequent sight adjustment it is necessary only to boresight as outlined in paragraph 32, unlock the boresight knobs, index the established zero settings, and lock the boresight knobs.

54

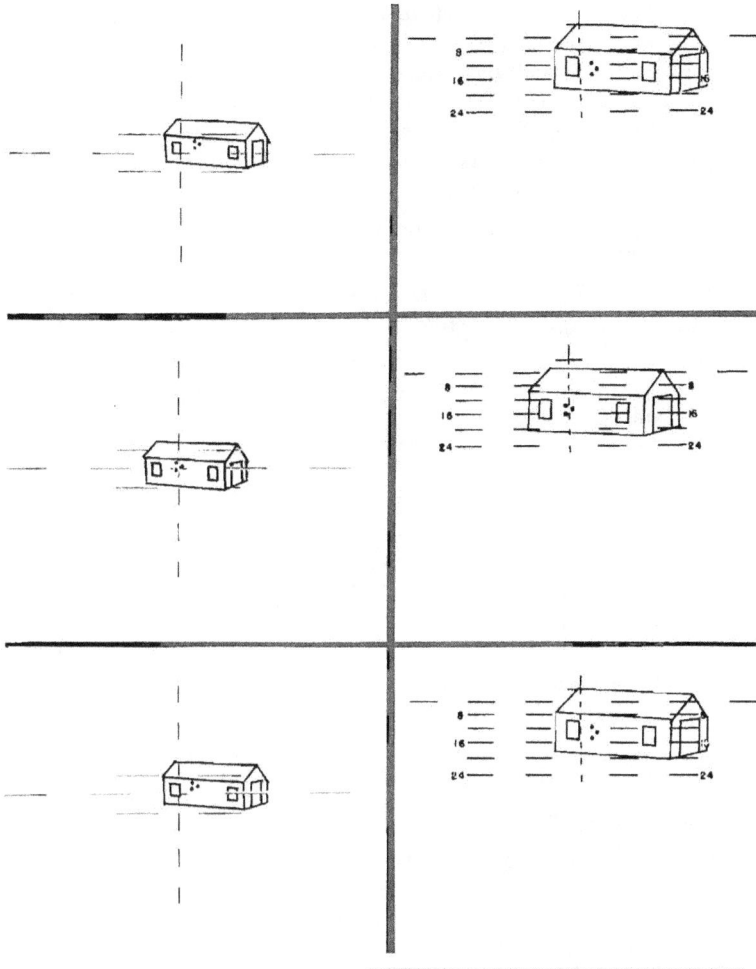

Figure 32. Zeroing.

Note. For full realization of first-round hit potential, great care must be taken in boresighting and zeroing. Check rounds must be fired periodically at definitely known ranges, and rezeroing must be accomplished when required. The accurate recording of all zeroing and check firing data in the gun book is essential to obtaining first-round hits as well as for conservation of ammunition.

34. Synchronization and Backlash

a. Synchronization. Using units are not authorized to adjust the linkage arms between the gun trunnion, ballistic drive, periscope, and range finder. They may check the linkage adjustment as follows:

(1) Boresight on a distant target.

(2) Place the tank on a steep forward slope, and check the boresight when the gun is near maximum elevation.

(3) Place the tank on a steep reverse slope, and check the boresight when the gun is near maximum depression.

(4) If the range finder or periscope is not accurate within .5 mil with the gun at maximum elevation or depression, the linkage should be adjusted.

Note. The actual adjustment of the linkage must be performed by Ordnance personnel.

b. Elevation Backlash. An elevation backlash check of the primary direct-fire sighting system may be performed as follows:

(1) Elevate the gun approximately 20 mils above a distant aiming point. Depress the gun until the aiming cross is on the aiming point, being careful not to pass the aiming point. Measure the existing gun elevation with the M1 gunner's quadrant.

(2) Depress the gun approximately 20 mils below the aiming point. Elevate until the aiming cross is again alined precisely on the aiming point, being careful not to pass the aiming point. Measure the existing elevation with the gunner's quadrant.

(3) The difference between the two quadrant readings is the elevation backlash in the system. If the backlash exceeds .3 mil, notify supporting Ordnance.

Note. The effects of backlash can be largely eliminated if the final lay for elevation is always made in the same direction.

35. Training of Stereoscope Range Finder Operators

a. General. The range finder is a precision optical instrument. In order to get the best results from this instrument, all phases of training must emphasize this point. The time required for a trained tank commander to range accurately will not exceed 5 seconds. In ranging, accuracy cannot be sacrificed for speed; both must be developed simultaneously.

b. Equipment and Areas Necessary for Training.

(1) The following throwover charts or similar visual aids are needed:

(a) Complete fire-control system, including linkage.

(b) Rear view of T46E1 range finder.

(c) Blow-up of T46E1 range finder control panel.

(d) Three views of range finder reticle patterns showing both eyepieces with stereo switch on:

1. ON-STEREO.

2. OFF.

3. ON-AUX. GUNSIGHT.

(e) Three stereo composites showing the ranging reticle suspended over the target with range set:
 1. Short.
 2. Over.
 3. At target range.

(2) The training area for range finder instruction must include a target-ranging area with ranges up to at least 3,500 yards. Targets should be placed in this area at known ranges from 700 yards to the maximum range available. There should be between 12 and 24 targets in all available types of locations, such as on forward slopes (steep and gradual), on the skyline, partially visible over a ridge line, against contrasting background, in thick vegetation, and in sparse vegetation. Some of the targets should be partially camouflaged. For preliminary training, targets with sharp outlines and little or no depth are desirable. As skill in ranging is gained, the tank commander can range on the more difficult targets accurately.

c. *Training Procedures and Techniques.*

(1) Physical examination of personnel prior to inauguration of range finder training is not necessary, but it is an aid in eliminating those who are physically incapable of operating a stereoscopic range finder. This is of particular importance in units that are short of equipment. The examination should determine whether or not personnel have stereoscopic vision. The examination should also include a measurement of the interpupillary distance for each man. The interpupillary adjustment on the T46E1 range finder will accommodate measurements from 58 to 72 millimeters.

(2) If physical examination of personnel by a medical officer is not possible, men falling into the following categories should be eliminated prior to training:
 (a) Those who have vision in only one eye.
 (b) Those who suffer serious muscular impairment of one eye.
 (c) Those whose interpupillary distance is less than 58 millimeters or greater than 72 millimeters.
 (d) Those who demonstrate physical, mental, or psychological deficiencies of a nature to render them incapable of performing the duties of a tank crewman.

(3) Use, operation, adjustment, and maintenance should be explained by use of conference and demonstrations. Particular emphasis will be placed on the following: diopter scales, interpupillary scale, halving knob, boresight adjustment knobs with slip scales, ICS knob, light switches and rheostat, range knob, filter lever, reticle lamps, and spare lamps.

(4) During the application phase, men must be checked individually to make sure they can see the ranging reticle in depth. It may be necessary to cross-examine the student, even though he claims he can see the reticle in depth.

(5) After the operator can see the ranging reticle in depth, continued practice will develop the required speed and accuracy. During early stages of training, the average man should not range for more than 20 minutes per hour. Optimum progress will be obtained if the ranging practice is given in 4-hour blocks. There should be a maximum of four men assigned to each tank, two of whom may be engaged in concurrent training. When the operator has completed ranging on a target, a second man, standing to his left, can read the range scale and record the indicated range. The operator ranges once on each target in turn. As proficiency increases, he should obtain an increasing number of readings per period.

(6) Initially, the operator should be allowed 30 seconds to complete one ranging. Ranging time should be reduced gradually until he consistently ranges, with accuracy, in five seconds or less.

(7) The operator must be taught to move the ranging reticle toward the target until the target appears to be bracketed in range between the lowest vertical bar of the reticle and the two center bars; final adjustment is made by moving the reticle until the lowest bar appears to be at the same range as the target.

(8) A new block of targets should be assigned each hour. Radio control may be used to control ranging periods, to rotate gunners, to designate blocks of targets to be used, and to notify personnel of breaks.

(9) Practice readings must be recorded to determine each operator's progress. Accuracy and speed increase as the number of rangings increases. Although some personnel will qualify after a few hundred rangings, it normally takes 1,800 readings for an average man to attain proficiency.

(10) During initial periods of ranging, the ICS knob is set at 25. The ranges indicated may vary considerably from true range. When the operator becomes consistent, his ICS setting will be determined, at which time his indicated range should approximate true range. To determine his individual ICS setting, the operator selects a target of known range as near 1,500 yards as possible, and indexes this range on the range scale of the range finder. He then rotates the ICS knob, causing the ranging reticle to move in depth until the lower

bar of the ranging reticle appears to be at the same range as the target. He notes and records the ICS scale reading. This procedure of ranging with the ICS knob is repeated until at least ten settings are recorded. The average of these settings is his personal ICS setting for the range finder on which it was determined. For maximum accuracy during subsequent ranging practice, the same range finder, with the ICS setting indexed, should be used. Should it become necessary to assign an individual to another range finder, his ICS setting will be indexed until such time as a new setting is determined. The setting may vary slightly from day to day; therefore the operator should check his ICS setting frequently.

(11) The training of range finder operators must be directed toward reducing the deviation in ranging to four units of error (UOE). The accuracy potential of the range finder can be fully exploited only when the operator consistently ranges within the allowable error at all ranges. The following chart shows the allowable four units of error for ranges from 500 to 4,800 yards.

Range	4 UOE (yards)	Range	4 UOE (yards)
500	3	1,600	27
600	4	1,700	31
700	5	1,800	34
800	7	1,900	38
900	9	2,000	42
1,000	11	2,500	66
1,100	13	3,000	95
1,200	15	3,500	130
1,300	18	4,000	170
1,400	21	4,500	215
1,500	24		

This table was compiled using the formula $10.6\dfrac{R}{(1000)^2}=4$ UOE, with R being the range in yards, and the result obtained taken to the nearest whole yard. For ranges other than those listed, an interpolation may be made for an approximate allowable error; or the above formula may be used for more accurate compilations. At any range, 4 UOE represents the range deviation, in yards plus or minus, from the given range. For example, if a man is ranging at 1,500 yards with his personal ICS indexed, his indicated range must fall between 1,476 (1,500−24) yards and (1,500+24) yards.

36. Vision Devices

The driver in the M48 tank has plain-window or plastic Periscopes, T25 or T26, for observation. In addition, the tank commander's cupola is provided with vision blocks or Periscope, M17, to permit all-around vision with the hatch closed.

37. Auxiliary Fire-Control Equipment

The Tank, 90–mm Gun, M48, is equipped with fire-control equipment for indirect laying of the tank gun when the target is not visible to the gunner. This equipment includes the Gunner's Quadrant, M1; Elevation Quadrant, M13; and Azimuth Indicator, T28.

38. Gunner's Quadrant, M1

a. General. The Gunner's Quadrant, M1 (fig. 33), standard equipment for each tank, is used to check the Elevation Quadrant, M13; to lay the gun for elevation when firing at point targets where greater accuracy is needed than can be obtained with the Elevation Quadrant, M13; and to measure elevation, depression, and cant of the gun.

b. Description.

(1) The Gunner's Quadrant, M1, consists of a sector-shaped frame to which is pivoted an index arm and level-vial holder. A scale graduated from 0 to 800 mils, at 10-mil intervals, is on one side of the frame, and a scale graduated from 800 to 1,600 mils is on the opposite side.

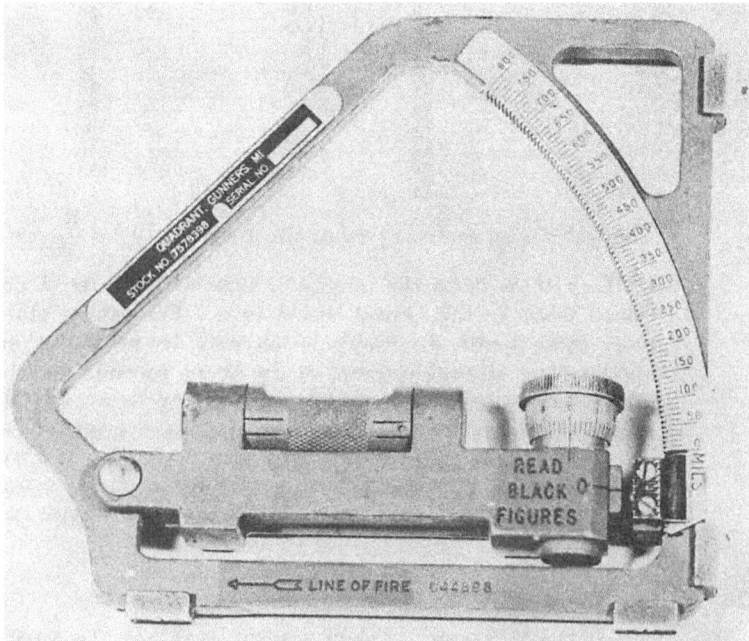

Figure 33. Quadrant, Gunner's, M1.

(2) The inside sector of the frame has teeth at 10-mil intervals. The teeth engage a spring-loaded, saw-toothed plunger in the index arm and thereby permit setting of the arm to the desired angle as indicated on the scale.

(3) A level and a micrometer mechanism are mounted on the index arm. The micrometer scale is graduted from 0 to 10 mils in .2-mil increments. The micrometer scale has red and black figures. The black figures are used when taking readings on the 0–800-mil scale, while the red figures are used with the 800–1,600-mil scale. For tank gunnery purposes, the black figures only are used.

(4) Auxiliary indexes on the index arm and on the plunger indicate whether the micrometer is to be read as 0 mils or as 10 mils. A zero micrometer indication is read as 0 mils when the auxiliary indexes are matched, and as 10 mils when they are not matched.

(5) Steel shoes are screwed to the frame on two sides to serve as two sets of true bearing surfaces for the quadrant.

(6) An arrow, with the instruction "LINE OF FIRE," is on each side of the frame to indicate the direction in which the quadrant should be pointed for normal use.

c. *Test and Adjustment.* Test of zero setting (end-for-end test) is accomplished as follows:

(1) Set both the index arm and the micrometer scale at 0.

(2) Place the quadrant on the quadrant seats of the breech ring, and center the bubble by elevating or depressing the gun.

(3) Turn the quadrant end for end. If the bubble recenters itself, the quadrant is in perfect adjustment. If the bubble does not recenter itself, try to center the bubble by turning the micrometer knot.

(4) If the bubble recenters the correction is plus (positive) and equal to one-half the micrometer reading. When laying the gun to a given elevation add the correction to the given angle. When measuring existing elevation angles subtract the correction from the micrometer knob reading.

(5) If the bubble does not recenter when the micrometer is turned the correction is minus (negative). The amount of correction is determined as follows: Drop the elevation index to −10 (one notch below zero); rotate the micrometer knob until the bubble is centered; subtract the micrometer reading from 10, and divide the remainder by 2. When laying the gun to a given elevation subtract the correction from the given elevation angle. In the event the remainder thus obtained is less than zero, drop the index to −10, subtract this remainder from 10, and index the resultant angle on the

micrometer. When measuring an existing elevation angle add the correction to the micrometer reading.

39. Elevation Quadrant, M13

a. The Elevation Quadrant, M13 (fig. 34), is mounted on the center of the ballistic drive elevation shaft. It is used to measure vertical angles and to provide a means of laying the gun for elevation.

b. The micrometer scale is graduated in both directions from 0 to 100 mils in 1-mil increments and is numbered every 10 mils from 0 to 90. The black figures are used for plus elevations and the red figures for minus elevations. One complete revolution of the micrometer knob moves the elevation scale index 100 mils on the elevation scale.

c. The elevation scale is graduated from −200 to +600 mils. It is marked every 100 mils and numbered every 200 mils from 0 to −2 and from 0 to +6.

d. The reflector and the level vial move with the elevation scale index when angles are set into the instrument. When the bubble is centered in the level vial, the gun is laid at the angle of elevation or

Figure 34. Quadrant, Elevation, M13.

depression set on the instrument. The reflector over the level vial provides a means for the gunner to see the bubble in the level vial.

e. Illumination of the level vial, elevation scale, and micrometer scale is provided by the Instrument Light, T22. The light switch, located to the right of the M13 elevation quadrant, has three positions: center—off; up—3-volt power source; down—24-volt power source. The up position is used for emergency operation when the tank's 24-volt system is not in operation.

f. The adjustment of the M13 elevation quadrant can be checked by means of the M1 gunner's quadrant. Level the gun using a corrected M1 gunner's quadrant. Without disturbing the lay of the gun, center the bubble in the level vial of the M13 quadrant by rotating the micrometer knob. Check the elevation scale. If 0 is not indexed on this scale, loosen the screw at each end of the scale and slip it until 0 is opposite the elevation scale index. Tighten the screws. Check the micrometer scale. If 0 is not indexed, loosen the three screws on the micrometer knob, then slip the micrometer scale. Check the bubble to be sure it is still centered in the level vial; if it is, tighten the three screws on the micrometer knob, and the instrument is ready for use. If the bubble is not centered, repeat the adjustment operation.

40. Azimuth Indicator, T28

a. The Azimuth Indicator, T28 (fig. 35), which is mounted on the right side of the turret with gears in mesh with the turret ring, measures horizontal angles of traverse. It is used principally for laying the gun for indirect fire. The azimuth indicator is a dialed instrument with pointers indicating the readings on the scales.

b. The Azimuth Indicator, T28, has three scales and three pointers. The azimuth scale is graduated in 100-mil intervals and is numbered every 200 mils from 0 to 3,200 counterclockwise in two consecutive semicircles around the scale. The micrometer scale is graduated counterclockwise in 1-mil intervals and numbered every 5 mils from 0 to 100. The gunner's aid is graduated in 1-mil intervals and numbered every 5 mils from 0 to 50 mils right and left. The directional pointer is fixed in relation to the longitudinal axis of the tank and gives a directional reading of the azimuth scale. This reading indicates the number of mils the gun has traversed from the longitudinal axis of the hull. The azimuth pointer works in conjunction with the micrometer pointer. These pointers are adjustable and may be set at zero (by depressing and turning the resetter knob) when the sights are set on any desired reference point. The sum of these two pointer readings then accurately indicates the number of mils the gun has traversed from the reference point. The azimuth and micrometer scales are fixed, while the gunner's aid may be rotated to any position. The gunner's aid is used to make shifts in deflection by rotating it

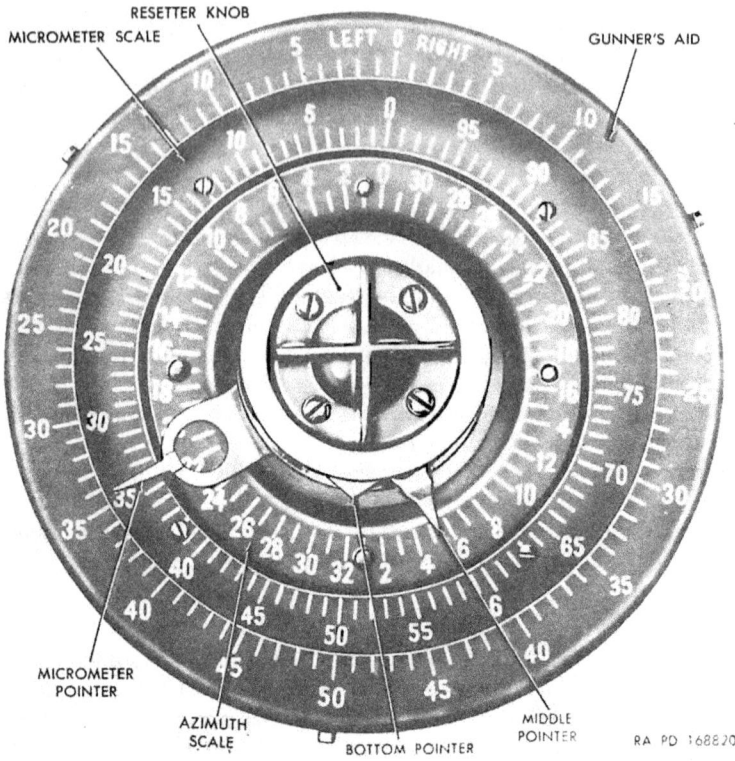

RESETTER KNOB
MICROMETER SCALE
GUNNER'S AID
MICROMETER POINTER
AZIMUTH SCALE
BOTTOM POINTER
MIDDLE POINTER
RA PD 168820

Figure 35. Azimuth Indicator, T28.

until the zero coincides with the position of the micrometer pointer. Right and left deflection corrections are then laid off on the gunner's aid. After the shift has been applied, move the zero of the gunner's aid to the micrometer pointer.

c. Built-in electric lamps provide illumination for the scales of the azimuth indicator. A receptacle on the side of the azimuth indicator receives the plug on the instrument light, which is installed in a bracket immediately above the indicator. The lamps, which are in the indicator, are turned on and off by a toggle switch on the instrument light.

d. To test the accuracy of the azimuth indicator, lay the aiming cross of the periscope on a definite aiming point, and set the azimuth and micrometer pointers at zero. Traverse the turret manually through a complete circle until the sight is laid back on the original aiming point. If the azimuth and micrometer pointers do not indicate zero, the azimuth indicator is out of adjustment and requires repair by Ordnance personnel.

e. To test the azimuth indicator for slippage, lay the aiming cross of the sight on a definite aiming point, and set the azimuth and micrometer pointers at zero. Traverse the turret rapidly in power, and stop suddenly; repeat this operation two or more times. Manually traverse the turret in the opposite direction back to the aiming point. If the azimuth and micrometer pointers do not indicate zero, the azimuth indicator is slipping and will require repair by Ordnance personnel. If the pointers indicate zero, repeat this check in the opposite direction.

f. To maintain the azimuth indicator, keep it clean and covered when not in use. Any lubrication or adjustment must be done by Ordnance personnel.

41. Instrument Lights

a. Instrument Light, M36. The Instrument Light, M36 (fig. 36), is a self-contained lighting device consisting of a tube, two flashlight batteries, a rheostat, and a 3-volt electric lamp. The batteries are contained in the tube and are held in position by a bayonet-type cap and spring which fits over one end of the tube. The rheostat is mounted in the other end of the tube. Rotate the rheostat knob to turn the electric power on and off and to regulate the intensity of the illumination. The electric lamp is mounted in a lead wire body which is screwed to the lamp bracket. The lamp bracket has a dovetail-formed base which contains a window and detent device. To position

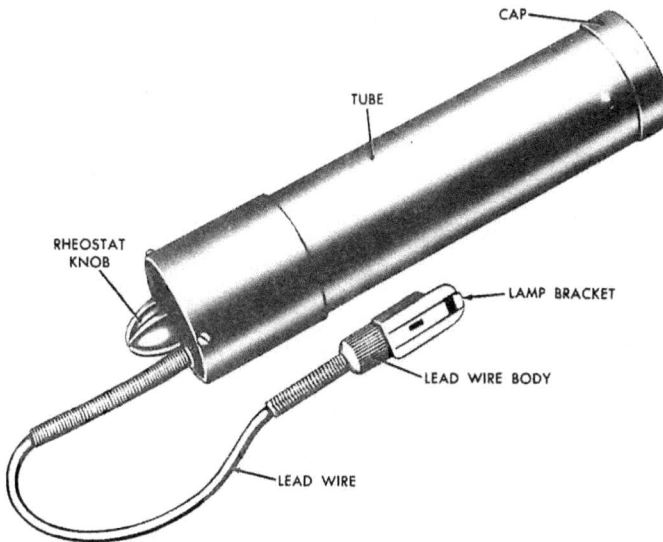

Figure 36. Instrument Light, M36.

the bracket on the telescope, slide it into a corresponding slot over the reticle until the detent is engaged. A block or slide on the side of the body provides for holding the lamp bracket when it is not attached to the telescope.

b. Instrument Light, D78454. The electrical power source for the built-in lamps of the Azimuth Indicator, T28, is provided by the Instrument Light, D78454, which is installed in a bracket immediately above the indicator. A plug on the lead wire assembly fits into a receptacle on the side of the azimuth indicator. The lamps which are in the indicator are turned on and off by a toggle switch on the instrument light.

Note. Instrument lights should be installed in the brackets at all times. Batteries should be installed only when the lights are in use.

CHAPTER 3

CREW DRILL AND SERVICE OF THE PIECE

Section I. INTRODUCTION

42. General

This chapter is primarily for the guidance of platoon leaders and tank commanders in training crew members with the objective of attaining efficient teamwork in the crew operation of the Tank, 90–mm Gun, M48. It is emphasized that the drills described in this chapter are for the development of crew teamwork in the fighting operation of the tank, and that the ultimate goal is successful operation of the tank on the battlefield.

Section II. CREW COMPOSITIONS AND FORMATIONS

43. Crew Composition

The crew of the Tank, 90–mm Gun, M48, consists of four members— tank commander, gunner, driver, and loader.

44. Formations

a. Dismounted Posts. The crew form in one rank, with the tank commander two yards in front of the right track. The gunner, driver, and loader take posts in line with, and to the left of, the tank commander, at close interval.

b. Mounted Posts. The crew form, mounted as follows:

(1) *Tank commander.* In the turret, standing on the tank commander's platform or seated on the tank commander's seat.

(2) *Gunner.* In the gunner's seat to the right of the tank gun and in front of the tank commander.

(3) *Driver.* In the driver's seat in the front center of the hull.

(4) *Loader.* On the left side of the tank gun, standing on the turret floor or seated on the loader's seat at the left rear of the turret.

Section III. CREW CONTROL

45. Operation of Interphone and Radio

The tank interphone system is used for voice communication between members of the tank crew, and for communication with individuals outside the tank through the external interphone. The tank radio set is used for communication with other tanks and with other units. The interphone is a part of the vehicular radio set. The equipment is designed so that operation of the interphone system will override received or transmitted signals, but will not cut transmitted signals off the air. The crew must be proficient in the operation of the interphone system if they are to obtain its maximum value in combat. Proficiency in the operation of the interphone system is gained only by continued practice.

46. Control Box Positions

a. Interphone control box positions are as follows:
(1) The tank commander and gunner plug into a single box located on the right wall of the turret.
(2) The loader plugs into a control box on the left wall of the turret.
(3) The driver plugs into a control box located to his right or right rear.

b. External interphone control box positions are as follows:
(1) External interphone OFF-ON switch installed in the driver's compartment.
(2) External interphone PUSH-TO-SIGNAL control installed in the vehicle turret in the vicinity of the loader's position.
(3) Cable reel consisting of 40 feet of cable terminating in a handset located in a compartment at the rear of the tank hull.

47. Modes of Operation

a. General. When power has been supplied to Set 1, Set 2, and the auxiliary receiver, and after squelch adjustments have been made, the following modes of operation are possible at each interphone control box:
(1) Monitoring of Set 1, Set 2, and the auxiliary receiver.
(2) Push-to-talk operation of Set 1 or Set 2.
(3) Interphone facilities between interphone boxes.

b. Monitoring. Monitoring received signals is accomplished by placing the selector switch pointer of the interphone control box in the center position. This position permits monitoring Set 1, Set 2, and the interphone in the Radio Set, AN/GRC–4, –6, and –8. It also permits monitoring the auxiliary receiver in the Radio Set, AN/GRC–3, –5, and –7.

c. Interphone Operation.
(1) *Tank interphone operation.* Interphone reception is possible with the selector switch in any position. To communicate with a crew member at any interphone box, press either the HOLD ON or the LOCK ON button on the chest set and talk into the microphone. In an emergency, any crew member can override a radio conversation without waiting for the sending party to stop talking.

(2) *External interphone operation.* To contact troops outside the tank, the driver turns on the external interphone and the loader operates the external interphone PUSH-TO-SIGNAL switch, thereby illuminating the external call light located on the bottom of the external interphone compartment. This light can be blinked by momentary operation of the switch. Troops outside the tank, responding to this signal, talk to the tank crew by removing the handset from the compartment, pressing the PUSH-TO-TALK-AND-LISTEN switch, and speaking into the handset over the tank interphone system. Volume level may be adjusted by turning the volume control in the external interphone compartment. Upon completion of the conversation, the handset and any unreeled cable is returned to the compartment and its lid closed and latched. To contact the tank crew from the outside, with the EXTERNAL-INTERPHONE ON-OFF switch in the ON position, remove the handset from the rear compartment and press the control switch. This illuminates the interphone control light, and communication with the crew can be established.

d. Radio Operation of Sets 1 and 2.

(1) For push-to-talk operation of Set 1, turn the selector switch pointer to the left-hand position, press the HOLD ON button and the RADIO button on the chest set, place the RADIO TRANS switch on the control box in the TRANS position, and talk into the microphone. Release the chest set buttons to listen. If the auxiliary receiver is not being used as a monitor station, and it interferes with operation of Set 1, turn the receiver VOLUME control to the OFF position.

(2) For push-to-talk operation of Set 2, turn the selector switch pointer to the right-hand position, press the HOLD ON button and the RADIO button on the chest set, place the RADIO TRANS switch on the control box in the TRANS position, and talk into the microphone. Release the chest set buttons to listen. When Set 2 is used, the loader is designated as monitor-operator.

48. Radio and Interphone Check

a. Inspections of communication equipment will be performed as prescribed on DA Form 11–238.

b. It is the duty of each crew member to check his interphone equipment. He should see that it is complete, in good working order, clean, and properly maintained. Any difficulties should be reported to the tank commander.

49. Use of Definite Terminology

Terminology prescribed for tank commanders in controlling their crews is set forth in the following paragraph. Failure to use standard, specific interphone language causes misunderstanding and disorder. Adherence by all crew members to this standard language is essential to efficient operation of the tank.

50. Interphone Language

a. Terms.

Tank commander_____ TANK COMMANDER.
Driver_____ DRIVER.
Gunner_____ GUNNER.
Loader_____ LOADER.
Any tank_____ TANK.
Any unarmored vehicle_____ TRUCK.
Any antitank gun or artillery piece_____ ANTITANK.
Infantry_____ TROOPS.
Machinegun_____ MACHINEGUN.
Airplane_____ PLANE.
Any other target_____ Briefest descriptive word or phrase.

b. Commands for Movement of Tank.

To move forward_____ DRIVER MOVE OUT.
To halt_____ DRIVER STOP.
To reverse_____ DRIVER REVERSE.
To increase speed_____ DRIVER SPEED UP.
To decrease speed_____ DRIVER SLOW DOWN.
To turn right (left)_____ DRIVER RIGHT (LEFT) STEADY . . . ON.
To pivot right (left)_____ DRIVER PIVOT RIGHT (LEFT) STEADY . . . ON.
To move toward a terrain feature or reference point, tank being headed in proper direction. — DRIVER MARCH ON WHITE HOUSE (HILL, DEAD TREE, ETC.).
To follow road or trail to the right (left). — DRIVER RIGHT (LEFT) ON ROAD (TRAIL).
To follow in column_____ DRIVER FOLLOW THAT TANK (DRIVER FOLLOW TANK B-9).
To start engine_____ DRIVER TURN IT OVER.
To stop engine_____ DRIVER CUT ENGINE.
To proceed in a specific transmission range. — DRIVER LOW (HIGH) RANGE.
To proceed at same speed_____ DRIVER STEADY.

c. Commands for Control of Turret.

To traverse turret_____ GUNNER TRAVERSE RIGHT (LEFT).
To stop turret traverse_____ STEADY . . . ON.

d. Fire Commands. See chapter 4.

Section IV. CREW DRILL

51. Dismounted Drill

a. To Form Tank Crew. Being dismounted, the crew take dismounted posts at the command FALL IN.

b. Fall In. On command, the crew fall in at attention. The tank commander takes his post two yards in front of the right track, facing to the front. The gunner, driver, and loader, in that order, take posts to the left of the tank commander at close interval.

c. To Break Ranks. Being at dismounted posts, the crew break ranks at the command of FALL OUT. Crew members habitually fall out to the right of the tank.

d. To Call Off. Being at dismounted posts, at the command CALL OFF, the members of the crew call off in turn as follows:

Tank commander_____ TANK COMMANDER.
Gunner_____ GUNNER.
Driver_____ DRIVER.
Loader_____ LOADER.

e. To Change Designations and Duties.

(1) The crew being at dismounted posts, at the command TANK COMMANDER (GUNNER) (DRIVER) FALL OUT:

(*a*) The man designated to fall out moves along the rear of the rank to the left flank position and becomes loader.

(*b*) The crew members on the left of the vacated post move one position to the right and prepare to call off their new assignments.

(*c*) The acting tank commander starts calling off as soon as the crew is re-formed in line.

(2) The movement may be executed by having any member of the crew fall out except the loader.

(3) All movements should be executed at double time with snap and precision.

52. To Mount the Tank Crew

This drill starts with the crew at dismounted posts.

Note. All phases of crew drill begin with the tank gun forward, positioned over the right edge of the driver's hatch, and the caliber .50 machinegun positioned to the front.

Tank commander	Gunner	Driver	Loader
Command: PREPARE TO MOUNT.			
About face	About face	About face	About face
Command: MOUNT.			
Stand fast		Stand fast	
Mount right fender	Mount right fender		Mount left fender
Mount right sponson	Mount right sponson	Mount left fender	Mount left sponson
Enter turret	Enter turret and take post	Enter driver's hatch	Enter turret and take post
		Turn on master switch	Turn on radio
Connect break-away plugs	Connect break-away plugs	Connect break-away plugs	Connect break-away plugs
Command: REPORT	Report: GUNNER READY	Report: DRIVER READY	Report: LOADER READY

53. To Close and Open Hatches

a. To Close Hatches. This drill starts with the crew at mounted posts.

Tank commander	Gunner	Driver	Loader
Command: CLOSE HATCHES.			
Close hatch	Close hatch	Close hatch	Close hatch.
Command: REPORT	Report: GUNNER READY	Report: DRIVER READY	Report: LOADER READY.

b. To Open Hatches. Crew being at mounted posts.

Tank commander	Gunner	Driver	Loader
Command: OPEN HATCHES.			
Open hatch		Open hatch	Open hatch.
Command: REPORT	Report: GUNNER READY	Report: DRIVER READY	Report: LOADER READY.

54. To Dismount Tank Crew

This drill starts with the crew at mounted posts, hatches open, and the tank gun forward and positioned over the right edge of the driver's hatch.

Tank commander	Gunner	Driver	Loader
Command: PREPARE TO DISMOUNT . . . DISMOUNT.			
Disconnect breakaway plugs	Disconnect breakaway plugs	Turn off master relay switch. Disconnect breakaway plugs.	Turn off radio. Disconnect breakaway plugs.
Emerge from turret	Remain in position	Emerge from hatch	Emerge from turret.
Move to right sponson	Emerge from turret	Move to left fender	Move to left sponson.
Move to right fender	Move to right sponson	Take dismounted post	Move to left fender.
Take dismounted post and command: CALL OFF.	Move to right fender. Take dismounted post.		Take dismounted post.

55. To Dismount Through Escape Hatch

Crew is at mounted posts.

Tank commander	Gunner	Driver	Loader
Command: THROUGH ESCAPE HATCH, PREPARE TO DISMOUNT . . . DISMOUNT.			
Disconnect breakaway plugs.	Traverse turret to give access to driver's compartment. Disconnect breakway plugs.	Turn off master relay switch. Disconnect breakaway plugs. Position driver's seat. Release escape hatch.	Turn off radio. Disconnect breakaway plugs.
Secure carbine and pass to driver.	Move to left side of turret.		Secure SMG.
Move to left side of turret.	Stand fast.		
Stand fast.	Dismount through escape hatch. Take dismounted post.	Dismount through escape hatch with carbine.	Dismount through escape hatch with SMG. Take dismounted post.
Dismount through escape hatch. Take dismounted post and command: CALL OFF.		Take dismounted post.	

56. Pep Drill

To vary the drill routine and to maintain the interest of the crew members, unexpected periods of pep drill are introduced into the training. Pep drill consists of a series of precision movements executed at high speed and terminating at the position of attention, either mounted or dismounted. For example, the crews being dismounted, the platoon leader commands: IN FRONT OF YOUR TANKS . . . FALL IN; MOUNT; DISMOUNT; FALL OUT TANK COMMANDER; ON THE LEFT OF YOUR TANKS . . . . FALL IN; FORWARD . . . MARCH; BY THE RIGHT FLANK . . . MARCH; TO THE REAR . . . MARCH; MOUNT. Preparatory commands for mounting and dismounting are eliminated in this drill. Posts of all crew members are changed frequently.

Section V. SERVICE OF THE PIECE

57. General

a. The gun crew in the tank consists of the tank commander, who ranges on targets, controls the fire, and, when necessary, adjusts the fire; the gunner, who aims and fires the tank gun and the coaxial machinegun; and the loader, who loads the guns.

b. Teamwork, coordination, precision, and speed are of utmost importance in service of the piece. Thorough training will provide a smooth, efficient operation in combat when speed is essential and delays or mistakes may be fatal.

58. Safety Precautions

a. Safety precautions and proper operating procedures are absolutely necessary if the tank is to be kept in operation. The procedures and precautions listed below should be repeated until the normal procedure is a *safe* procedure.

b. The loader will—

    (1) Check the breechblock crank stop to insure that it is in the locked position.

    (2) Check the bore of the gun for obstructions prior to and during firing.

    (3) Not allow the fuze or the primer of the round to strike any solid object in the turret.

    (4) Carefully examine each round of tank gun ammunition to see that it is clean and not bulged or dented.

    (5) Coordinate with the gunner prior to removing ammunition from hull stowage.

(6) Not attempt to disassemble any portion of a round of tank ammunition unless ordered to install a concrete-piercing fuze on HE ammunition.

(7) Stay clear of the path of recoil during and after loading the gun.

(8) Not attempt to trip the extractors with his fingers when closing the breech.

(9) Not remove the coaxial machinegun from the tank until it has been cleared and inspected by the tank commander.

c. The gunner will—

(1) Warn the crew before firing the main armament or coaxial machinegun. During training, he will pause one second after announcing ON THE WAY.

(2) Alert the crew before traversing the turret in power.

(3) Release the hand firing level after firing the gun manually, to avoid injury to the loader or damage to the gun as the next round is loaded.

(4) Turn off the 90–mm gun switch and announce MISFIRE when the main gun fails to fire.

(5) Turn off the coaxial machinegun switch and announce STOPPAGE when the coaxial machinegun fails to fire.

d. The tank commander must know and enforce all necessary safety precautions within his tank.

e. Any individual who observes a condition making firing unsafe will immediately call or signal the command CEASE FIRE.

59. To Open Breech

Grasp the grip portion of the operating handle, and pull it to the rear and down. *When the breech is locked open, immediately return the operating handle to its latched position.*

60. To Load Gun

a. Open the breech, and return the operating handle to its latched position (check engagement of latch).

b. Select a round of ammunition; grasp it by the base of the shell case with the right hand and by the rear of the ogive with the left hand.

c. Place the projectile in the loading notch, taking care not to strike the fuze. Move the round forward until the projectile rests in the chamber; remove the left hand, and push the round until the projectile is well into the chamber. Close and join the fingers of the right hand; then, with the heel of the right hand, vigorously push the round forward into the chamber, rotating the body to the left and sliding the hand off the round upward and to the left to insure clearing the breech. The breechblock will automatically push the hand clear if it should follow the round too far into the breech recess. Move to the left side of the turret, clear of the path of recoil, and announce UP.

61. To Unload an Unfired Round or a Misfire

a. To unload an unfired round, the loader cups his hands behind the breech to catch the base of the round as it emerges and to prevent it from dropping to the floor. The gunner, assisted by the tank commander, opens the breech *slowly.* (*The breech must not be opened rapidly, or the case may separate from the projectile.*) The loader then removes the round and returns it to its rack.

b. To unload a misfire, the following steps will be taken: Two more attempts, one electrically and one manually, will be made to fire the round. Wait one minute from the time of the last attempt before opening the breech. Insure that personnel unnecessary to the operation are cleared from the vicinity. Then remove the round. Rounds which misfire will not be returned to the racks, but will be removed to a safe place and turned over to Ordnance personnel.

62. To Remove a Stuck Round

When a round is stuck in the gun and it is impossible or inadvisable to fire it out, it will be removed. The loader attempts to remove the round with the extracting and ramming tool, placing the tips of the fork down and behind the rim of the stuck round and applying pressure. If this method fails, an attempt is made to remove the round using the bell rammer. The loader takes position to receive the round as it is pushed from the chamber. The tank commander dismounts, inserts the bell rammer into the muzzle of the gun, and pushes it gently down the bore until it is seated on the ogive of the projectile. Exerting a steady pressure, he shoves the round clear so that it may be removed by the loader. To the maximum possible extent, personnel should keep all parts of their bodies clear of the muzzle or breech during the operation. If this procedure fails to remove the round, Ordnance personnel will be called.

63. To Remove a Stuck Projectile

If the case and projectile become separated despite care in opening the breech, the chamber must be filled with rags to form a cushion to prevent ignition of a base-detonating fuze and to protect the breechblock. With the breech closed, the procedure described in paragraph 62 will be followed. After the projectile is free in the chamber, the breech will be opened and the projectile removed and disposed of in accordance with existing regulations. The chamber must be cleaned.

Section VI. MOUNTED ACTION

64. General

Prior to mounted action drill, the following conditions must be met:

a. Crew mounted.

b. Hatches open.

c. Tank gun forward.

d. Turret-mounted machinegun uncovered.

e. Ammunition stowed.

65. Prepare To Fire

A series of checks to turret components must be systematically performed by the tank crew to insure that the equipment is in proper working condition. These checks are performed before every operation. During training, the tank crew are drilled to perform these duties to insure completeness of checks, coordination of effort, and speed of execution. All checks listed must be performed in the assembly area. A final check is made in the attack position just prior to crossing the line of departure. Commands and duties of crewmen are listed below. The items marked by asterisks (*) must be included in the final check.

Tank commander	Gunner	Driver	Loader
Command: PREPARE TO FIRE.			
*Clean gunner's and tank commander's periscope and end windows of range finder.	Check recoil oil by physically feeling indicator tape. *Clean and inspect M20 periscope.	*Clean periscopes, lower seat, close hatch, turn on master switch.	*Open breech, inspect tube and chamber for obstructions and and cleanliness. Close breech.
*Check range finder and place it into operation.	*Clean and inspect telescope (interior). Check instrument lights and install batteries.		Check and adjust headspace on coaxial machinegun. Inspect all turret-stowed ammunition for completeness of stowage, type, and serviceability.
Command: CHECK FIRING SWITCHES	Turn 90-mm gun switch to ON position. Check firing trigger on power control handle.	Start auxiliary engine- - - -	*Turn manual safety to OFF position. Watch action of solenoid and listen for click of percussion mechanism after each firing check.
Recock main gun after each firing check. Check firing trigger on power control handle.	Check firing trigger on manual control handle. Check manual firing control.		

Check firing trigger on power control handle.	Turn off 90-mm gun switch and turn on coaxial machine gun switch.	Close cover and cock coaxial machinegun. Watch action of solenoid and listen for strike of firing pin.
	Check firing trigger on power control handle. Check firing trigger on manual elevation control handle.	Recock coaxial machinegun after each check.
Command: CHECK POWER CONTROL.	Turn off coaxial machinegun switch.	
	Check oil in turret power control system. *Unlock turret. Check manual traverse (to insure free movement of turret). *Check manual elevation. *Turn on turret motor switch.	*Check for obstruction to traverse.
*Check power control handle for power elevation and power traverse.	*Check power control handle for power elevation and power traverse.	

Tank commander	Gunner	Driver	Loader
Check and adjust headspace and timing on turret- or cupola-mounted caliber .50 machinegun. Check operation of mount and controls.	Check azimuth indicator for accuracy; traverse turret a complete rotation, stopping at one point to permit loader to check ammunition in hull stowage. Coordinate with crew members.		Check hull-stowed ammunition for completeness of stowage and serviceability; coordinate with gunner while checking ammunition.
	Place turret in power and check azimuth indicator for slippage. Turn off turret motor switch. Check elevation quadrant by use of M1 gunner's quadrant.		Open breech of main gun; return operating handle to latched position.
*Turn on computer switch and notify gunner.	Check manual operation of computer to insure no bind in computer or linkage. Push reset button on computer.		
Rotate range knob from 500 yards to 4,800 yards, then return to 500 yards, making intermittent stops at various ranges.	Observe to see that pointers on computer synchronize at various indexed ranges.		
	Index various types of ammunition and check for synchronization of pointers each time ammunition selector handle is released.		

Boresight range finder and apply established zero if necessary.

*Set unit battle sight on range finder.

*Half-load turret- or cupola-mounted caliber .50 machinegun. Turn off computer switch.

Command:
REPORT _____

Boresight M20 periscope and telescope and apply established zero to M20 periscope if necessary.

*Index ammunition on computer for unit battle sight. Await command to report.

Report:
GUNNER READY _____

Report:
DRIVER READY ____

*Half-load coaxial machinegun. Await command to report.

Report:
LOADER READY.

83

66. Gun Drill

A tank crew must be drilled in the performance of their firing duties to insure coordination of effort and speed of execution. Gun drill is conducted in the form of nonfiring exercises against both stationary and moving targets. Speed must be emphasized throughout this phase of drill. Gun drill must be periodically scheduled and conducted in order to maintain a high standard of tank crew proficiency. To stimulate interest, the tank should move a few yards between each nonfiring exercise, preferably over a simulated combat course in which there are various types of targets that become visible as the tank advances along the course. For the moving target phase of gun drill, a target mounted on a ¼-ton truck can be used. The speed and direction of travel of the target or target vehicle should be varied. The general firing duties of the crew are listed below. For specific firing duties in response to fire commands, see chapter 4.

Tank commander	Gunner	Driver	Loader
Be continually alert for targets. Control operation of tank by interphone.	Observe in assigned sector	Observe terrain for best routes and avoid obstacles when possible.	Observe in assigned sector.
Give fire commands; lay tank gun for direction and range on targets.	Index ammunition on computer; make final lay of gun; fire on targets and adjust fire for target destruction.	Observe in assigned sector.	Load ammunition announced in fire commands. Announce UP.
Observe fire and give subsequent fire commands, if gunner is unable to adjust fire.			Reload tank gun until CEASE FIRE is announced.
	Announce MISFIRE if tank gun fails to fire.		
Follow misfire procedure			Follow misfire procedure.
	Announce STOPPAGE if coaxial machinegun fails to fire.		Reduce stoppage; fire coaxial machinegun manually if so directed by gunner.
Fire turret-mounted machinegun as necessary.			Keep ammunition available in turret. Refill ready racks as necessary. Keep record of ammunition expended.

67. To Clear and Secure Guns

The clear-and-secure-guns procedure, like other procedures in tank operations, is conducted as a drill during training to insure that each crewman knows the duties he must perform in the clearing of the tank weapons and preparing the tank for an administrative move. If it is desired only to clear the tank weapons, the command is CLEAR GUNS. When the weapons are already cleared and it is desired to secure them, the command is SECURE GUNS. When it is desired to perform both phases together, the command is CLEAR AND SECURE GUNS. The crewmen's duties are listed below.

Tank commander	Gunner	Driver	Loader
Command: CLEAR AND SECURE GUNS.			
Clear turret- or cupola-mounted machinegun. Insert T-block. Turn off computer switch. Inspect and turn off range finder.	Turn off firing switches and turret motor switches. Inspect computer, periscope, and telescope.	Shut off auxiliary engine.	Clear coaxial machinegun. Insert H-block. Clear tank gun; inspect tube and close breech.
Place cover on turret- or cupola-mounted machinegun; secure gun in travel lock.	Coordinate with loader in rearrangement of ammunition stowage.		Fill ready racks.
Assist gunner in placing gun in travel lock.	Place gun in travel lock. Place cover on azimuth indicator. Assist loader in placing breech cover on tank gun; turn off instrument lights and remove batteries. Await command to report.	Place muzzle cover on tank gun. Await command to report.	Secure gun in travel lock. Place breech cover on tank gun. Await command to report.
Command: REPORT	Report: GUNNER READY	Report: DRIVER READY	Report: LOADER READY.

68. Loading All Weapons

The tank weapons are loaded on command. This is normally the fire command. The unit SOP may state the type of ammunition to be carried in the chamber of the tank gun before a target appears. The machineguns are half-loaded during PREPARE TO FIRE. In combat, the machineguns will be fully loaded when the unit is deployed for action.

69. Stowage and Handling of Ammunition

a. Ammunition stowage racks in the M48 tank are located on both sides of the driver's position in the hull, and readily accessible stowage racks are located in the turret. The hull stowage racks provide for stowage of 30 rounds, 19 on the left side of the driver and 11 on the right side. The turret stowage consists of a six-round stowage box located on the turret floor beneath the tank gun, a 16-round vertical stowage rack located along the turret ring, and an eight-round horizontal ready rack in the turret bustle. Ammunition should be stowed according to the unit SOP so that each type of ammunition is readily available to the loader.

b. Ammunition must be handled carefully to avoid striking the fuse or primer of the round against a hard surface. Each round must be inspected for dents and bulges and for dirt before stowing it in the tank. HE ammunition will be received and stowed with the fuze set at SQ (superquick). Because the primer is the most sensitive part of a round, ammunition should be passed from the ground into the tank with the primer up.

Section VII. DISMOUNTED ACTION

70. To Provide Local Security, Dismounted

When in assembly areas or under similar static conditions, it may be desirable to provide local dismounted security for the tank. A means of providing dismounted security, without detracting from the ability of the tank to go into action immediately, is for the tank commander to designate a crewman to dismount with hand grenades and carbine. Regardless of which crewman is designated, the remaining crewmen will occupy the positions of the driver, loader, and tank commander. The dismounted crewman will move to the position designated by the tank commander and provide local security.

71. To Abandon Tank

a. Ordinarily, the tank is abandoned as a result of a direct hit which either causes a fire that cannot be extinguished or disables the tank so that it becomes a vulnerable target. At the command ABANDON TANK, crew members open the hatches, dismount, and take cover a safe distance from the tank. The tank commander takes the carbine with him and covers the movement of his crew. The loader dismounts with the submachine gun. In case of fire, it is particularly important to hold the breath until clear of the vehicle; inhaling fumes and smoke may cause serious injury.

b. If it is necessary to abandon the tank, and if time permits deliberate action, the tank commander displays the flag signal DISRE-GARD MY MOVEMENTS (FM 21–60), and supervises the disabling of those weapons remaining in the tank. Back plates are removed from machine guns, and the firing pin and guide are removed from the tank gun. Like corresponding items in spare parts kits are also removed. Individual weapons and maximum ammunition loads are carried.

72. To Destroy Tank

When the command DESTROY TANK is given, crew members first remove all equipment that is to be carried. They then destroy the tank and the remaining weapons, ammunition, and equipment as prescribed by the unit SOP. (See sec. IX, this chapter.)

Section VIII. EVACUATION OF WOUNDED FROM TANKS

73. General

Wounded members of the tank crew normally will evacuate themselves from a disabled tank or be removed by their fellow crew members. The utmost speed is necessary in order to save the lives of those who are unhurt as well as the life of the casualty. A burning tank can trap the crew members in a matter of seconds; therefore it is essential that all crewmen become extremely proficient in utilizing the quickest methods of removing each other from the tank. If the action has ceased momentarily, or if the tank has been able to disengage itself without hindering the accomplishment of the mission, the casualty is removed and carried to a protected place, where emergency first aid is administered. Otherwise, the action will be continued until an opportunity is presented to remove casualties.

74. Methods of Evacuation

Methods of evacuation described herein are based on the employment of a two-man team, the largest team that can work effectively around a single hatch opening. In some cases, a third man will be able to give considerable help from the inside by placing belts around the wounded man or by moving him to a position where he can be grasped from above. The need for swift action usually will require that the casualty be grasped for removal by portions of his clothing or by the arms. If a limb is broken, or if there are other injuries which will be aggravated by these procedures, and if time allows, some form of lifting sling may be improvised to remove the crewman. Any equipment which is immediately available, such as pistol belts or web belts, may be used for this purpose.

75. Evacuation Drill, General

a. This paragraph contains general information which may be used as a guide in practicing the evacuation of crew members from any position. During drill, the composition of the evacuating team should be changed frequently to provide practice for all members of the crew in meeting various emergencies.

b. The member of the crew who discovers a wounded crewman announces DRIVER (LOADER, etc.) WOUNDED. If the tank is not actively engaged and the tank commander decides that evacuation is necessary, he commands: EVACUATE DRIVER (LOADER, etc.). Crew members dismount, unless one man is needed to help from inside, and the two nearest the hatch above the wounded man (No. 1 and No. 2 in pars. 76 and 77) take stations at the hatch to act as the evacuation crew. If the man nearest the casualty sees that his help is needed, he stays inside and arranges a sling or takes whatever steps he can to speed the operation. First aid is administered, and the wounded man is moved to a sheltered position. The tank commander reports the casualty.

76. Procedure To Evacuate Casualty From Driving Compartment

Tank commander commands—EVACUATE DRIVER. Loader unlocks driver's hatch from the inside. Tank commander opens the driver's hatch from the outside.

No. 1 (Tank commander)	No. 2 (Loader)
Take position on edge of hatch_____	Take position on edge of hatch.
Reach into hatch and grasp hands of casualty, straightening him in seat if necessary.	
Cross casualty's arms over his chest_____	Grasp nearest hand of casualty when his arms are crossed.
Rise and rotate casualty so that he faces to the rear.	Raise casualty and help rotate him.
Seat casualty on front rim of hatch; support him in this position while No. 2 jumps to ground.	Help seat casualty; jump to ground.

No. 1 (Tank commander)	No. 2 (Loader)
Lower trunk of casualty into arms of No. 2.	Receive and support trunk of wounded man, holding him beneath arms, around chest.
Lift legs out of hatch as No. 2 lowers casualty along slope plate.	Lower casualty along slope plate and support him until No. 1 can reach ground and assist.
Jump to ground; help No. 2 place casualty in carry position.	Place casualty in carry position.
Carry casualty to protected area_____	Help No. 1 carry casualty to protected area.

77. Procedure To Evacuate Casualty From Turret

This procedure can be followed in evacuating any turret crewman. The loader will be used in this example. The tank commander commands—EVACUATE LOADER (GUNNER, etc.), and dismounts to the rear deck, where he acts as No. 1. The driver acts as No. 2. The gunner, in this example assists by positioning straps or belts about the loader so as to obtain the maximum leverage. The gunner unlocks the loader's hatch and opens it with the aid of No. 1.

No. 1 (Tank commander)	No. 2 (Driver)
Take position on turret beside loader's hatch.	Raise casualty as high as possible in hatch opening, holding him around chest.
Grasp casualty under arms.	
Raise casualty through hatch, and seat him on rear edge.	Help No. 1 raise casualty by lifting from below.
Hold casualty while No. 2 dismounts to rear deck.	Dismount to rear deck.
Pick casualty up in arms; carry to rear and lay on back edge of deck.	Help No. 1 pick up casualty and carry to rear of tank; jump to ground.
Help No. 2 lift trunk of casualty off tank; jump to ground.	Lift upper part of casualty's body off tank and support until No. 1 arrives to help.
Lift casualty's hips and legs off tank.	
Carry casualty to protected area_____	Help carry casualty to protected area.

Section IX. DESTRUCTION OF EQUIPMENT

78. General

a. The destruction of materiel is a command decision to be carried out only on authority delegated by the division or higher commander. This usually is made a matter of standing operating procedure. *De-struction is ordered only after every possible measure for preservation or salvage of the materiel has been taken and when, in the judgment of the military commander concerned, such action is necessary to prevent:*

(1) Its capture intact by the enemy.

(2) Its use by the enemy, if captured, against our own or allied troops.

(3) Its abandonment in the combat zone.

(4) Knowledge of its existence, functioning, or exact specifications from reaching enemy intelligence agencies.

b. The principles followed are—

(1) Methods for the destruction of materiel subject to capture or abandonment in the combat zone must be adequate, uniform, and easily followed in the field.

(2) Destruction is as complete as possible within limitations of time, equipment, and personnel available. If thorough destruction cannot be completed, the most important features of the materiel are destroyed, and parts which cannot be easily duplicated and are essential to the operation or use of the materiel are ruined or destroyed. *The same essential parts are destroyed on all like units to prevent the enemy from constructing a complete unit from several damaged ones.*

c. Crews are trained in employing prescribed methods of destruction. *Training does not involve actual destruction of materiel.*

d. Certain methods of destruction require special tools and equipment, such as TNT and incendiary grenades, which may not be available. The issue of such special tools and equipment and the conditions under which destruction will be effected are command decisions, and depend upon the tactical situation.

e. The proper methods for destruction of the on-vehicle materiel and the tank are covered in TM 9-7012.

CHAPTER 4

CONDUCT OF FIRE

Section I. INTRODUCTION

79. General

The material contained in this chapter concerns the conduct of fire (direct fire) for the Tank, 90-mm Gun, M48, with Phase IV fire-control equipment. The proper and timely utilization of the excellent fire-control equipment will enable the M48 tank gun crew to obtain, with exceptional speed, a very high percentage of first-round hits. In order to realize the full capabilities of the M48 tank, the crew must be thoroughly drilled in their firing duties. This training will be conducted prior to any subcaliber or service firing and consists of nonfiring exercises for both stationary and moving targets. After qualification, the drill is conducted at frequent intervals to insure that a high level of crew proficiency is maintained. For basic principles of conduct of fire, see FM 17–12.

Section II. FIRING DUTIES

80. General

The effectiveness of the fire of tank weapons is entirely dependent on the coordinated action of the tank crew. Listed below are the general firing duties performed by the individual crew members.

Crew member	Firing duties
a. Driver	Moves the tank as directed by the tank commander. Observes for targets.
b. Loader	Loads the tank gun and coaxial machinegun; reduces stoppages; inspects, cleans, and stows ammunition; assists in removal of misfires, separated rounds, and stuck rounds; controls manual safety.
c. Gunner	Observes for targets; aims and fires the tank gun and coaxial machinegun by means of the periscope, ballistic computer, telescope, gun switches, traversing and elevating controls, and firing triggers; adjusts fire of tank gun and coaxial machinegun.

Crew member	Firing duties
d. Tank commander	Observes for and selects targets; controls movement of the tank and actions of the crew; gives initial fire commands and subsequent fire commands when necessary; using the tank commander's power control handle and range finder, lays gun initially for direction and ranges on the target; controls computer switch; supervises and assists gunner in adjusting fire; controls volume of fire, and fires caliber .50 turret- or cupola-mounted machinegun.

Section III. FIRING AT STATIONARY TARGETS

81. Initial Fire Commands and Firing Duties

The tank commander controls the fire of his tank and coordinates the actions of his crew by the timely issuance of fire commands. The initial fire command contains the necessary information for the crew to load, aim, and fire the tank weapons. Listed below are examples of these commands, together with the specific duties performed by the crewmen in response to the commands under the various conditions that might be encountered.

a. *Condition 1.*

(1) Primary fire-control equipment, consisting of range finder, ballistic computer, and M20 periscope.

(2) Target: stationary tank.

Note. In early stages of training, the tank commander will range on targets with the computer switch turned off; upon completion of ranging, he will turn the computer switch on. More speed can be attained in later stages of training with the computer switch turned on while the tank commander ranges; however, this will work effectively only with an experienced gunner and tank commander.

Element	Command	Tank Commander	Gunner	Loader
Alert	GUNNER	Using the range finder and the power control handle, lays the gun for direction and places the ranging reticle above or alongside the target, while announcing the first three elements of the initial fire command. Example: GUNNER, SHOT, TANK.	Turns on turret motor switch if turret is not in power.	Stands by.
Ammunition	SHOT		Using ammunition selector handle, indexes the announced ammunition on the computer. Turns 90-mm gun switch to ON position.	Selects and loads a round of shot, moves clear of the path of recoil, and announces UP. Selects another round of shot.
Range	Omitted [1]	Ranges on target; after completion of ranging, turns on computer and gives the remaining elements of the initial fire command. Example: FIRE.		

[1] If the range finder is damaged, the tank commander indexes his estimated range on the range finder and turns on the computer switch.

Element	Command	Crewmen's firing duties		
		Tank commander	Gunner	Loader
Direction	Omitted.[2]			
Target description	TANK		Announces IDENTIFIED when he sees the target.[3]	
Command to fire	FIRE		Takes control of the turret, makes final precise lay, announces ON THE WAY, and fires.	Loads round of shot and continues to load shot without command until CEASE FIRE or a change in ammunition is announced.

[2] If the power control is inoperative, the tank commander must give a direction element in the initial fire command. When the gunner identifies the target, he manipulates the gun so that the commander can range.

[3] If the computer switch is in the ON position, the gunner must manipulate his controls and coordinate with the tank commander so that the ranging reticle will be kept at the proper position for ranging.

b. Condition 2.

(1) Primary fire-control equipment, consisting of range finder, ballistic computer (operated manually), and M20 periscope.

(2) Target: Stationary truck.

| Element | Command | Crewmen's firing duties | | |
		Tank Commander	Gunner	Loader
Alert	GUNNER	Using the range finder and power control handle, lays the gun for direction and places the ranging reticle above or beside the target, while giving the first three elements of the initial fire command. Example: GUNNER, HE, TRUCK.	Turns on turret motor switch if not in power.	Stands by.
Ammunition	HE		Using the ammunition selector handle, indexes the announced ammunition on the computer. Turns 90-mm gun switch to ON position.	Selects and loads a round of HE, moves clear of the path of recoil, and announces UP.
Range	Omitted [1]	Ranges on target, then announces the remainder of initial fire command.[1] Example: FIRE.		Selects another round of HE.

[1] If the range finder is inoperable as a ranging instrument, the tank commander must index his estimated range on the range finder.

Element	Command	Crewmen's firing duties		
		Tank commander	Gunner	Loader
Direction	Omitted.²			
Target description	TRUCK		Announces IDENTIFIED as soon as he sees the target.	
Command to fire	FIRE		Uses the manual range crank on the computer, matches outer pointer index with inner pointer index; takes control of turret, makes final precise lay, announces ON THE WAY, and fires.	Loads round of HE and continues to load HE without command until CEASE FIRE or a change in ammunition is announced.

² If the power control is inoperative, the tank commander must give a direction element in the initial fire command.

c. *Condition 3.*

(1) Secondary fire-control equipment, consisting of the telescope, range finder, and ballistic computer.

(1) Target: Stationary heavy tank.

Element	Command	Crewmen's firing duties		
		Tank commander	Gunner	Leader
Alert	GUNNER..	Using the range finder and power control handle, lays gun for direction and places ranging reticle above or alongside the target, while giving first three elements of the initial fire command. Example: GUNNER, HYPER-SHOT, TANK.	Turns on turret motor switch if not in power.	Stands by.
Ammunition	HYPER-SHOT.		Indexes HYPER-SHOT on the computer. Turns 90-mm gun switch to ON position.	Selects and loads a round of HYPER-SHOT, moves clear of the path of recoil, and announces UP. Selects another round of HYPER-SHOT.

| Element | Command | Tank commander | Crewmen's mini duties | |
			Gunner	Leader
Range	Omitted	Ranges on the target, and announces the remainder of the initial fire command. Example: FIRE.	Manually cranks range into computer until outer pointer index matches inner pointer index. Then, using the ammunition selector handle, indexes APT–T33E7. The outer pointer will then indicate the range that the gunner must use on the telescope in order to hit a target with HVAP ammunition.[2]	
Direction	Omitted			
Target description	TANK		Announces IDENTIFIED as soon as he sees the target.	
Command to fire	FIRE		By manipulating his turret controls, lays appropriate range line on center of the target, announces ON THE WAY, and fires.	Loads round of HYPER-SHOT and continues to load HYPER-SHOT without command until CEASE FIRE or a change of ammunition is announced.

[1] If there is no power in the turret, the tank commander must estimate the range and index it on the range finder or announce it.

[2] If the computer cannot be operated manually, an aiming data chart must be used to determine the correct range line to use in the telescope.

82. Sensings

Rounds are sensed in relation to the target. The tank commander and gunner will mentally sense each round for range and deflection. These sensings are not announced unless the gunner fails to observe the burst or tracer through his direct-fire sight, in which case the gunner will announce LOST. The tank commander then will announce his sensing for range only, and his subsequent fire command. The five possible range sensings are—

a. TARGET. A round is sensed as TARGET only when the round is observed actually to strike the target, causing the target to change shape, pieces to fly off, or the target to completely disappear. When shot strikes a metal object, there is usually a distinctive orange flash.

b. OVER. A round is sensed as OVER when the burst appears beyond the target or the tracer passes above the target. This sensing is readily identified when firing HE, since the burst tends to silhouette the target.

c. SHORT. A round is sensed as SHORT when either the burst or the point of strike is observed between the gun and the target. The target is sometimes temporarily obscured by smoke and/or dust.

d. DOUBTFUL. A round is sensed as DOUBTFUL when the burst appears to be correct for range but off in deflection, or when the tracer passes to the right or left of the target but apparently is correct for range. A range change is not made on a DOUBTFUL sensing; a deflection correction normally is sufficient to secure a target hit.

e. LOST. A round is sensed as LOST when the tank commander or gunner fails to observe the point of strike, burst, or tracer. The point of strike may not be visible due to obscuration, terrain, or failure of the round to detonate. (Based on the tank commander's terrain appreciation, he may be justified in making a range change.)

83. Adjustment of Direct Fire, Primary Method

The primary method of adjustment is burst-on-target, in which the gunner, observing through his direct-fire sight, notes the point on the sight reticle where the burst or tracer appears in relation to the target and, without command from the tank commander, moves that point of the gun-laying reticle onto the center of the target before firing the next round. This method of adjustment provides a quick, accurate means of obtaining second-round target hits. The gunner uses this method whenever possible. Typical examples of burst-on-target adjustment, using the M20 periscope reticle as well as the telescope reticle, appear below.

a. Situation 1. Primary fire-control equipment, stationary target, shot ammunition.

(1) The gunner immediately re-laid after firing the first round so that the tracer appeared on his sight reticle in its proper relation to the target. This round was over and off in deflection to the left (fig. 37).

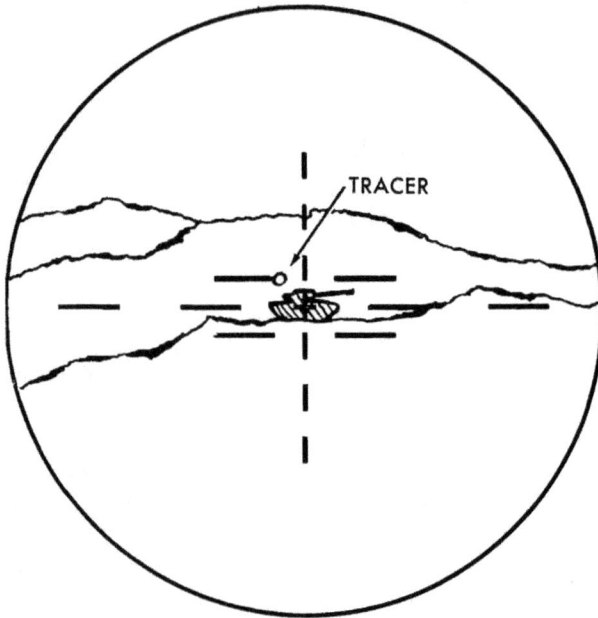

Figure 37. Situation 1. Stationary target.

(2) The gunner mentally noted the point on the sight reticle where the tracer appeared and, with the turret controls, moved that point to the center of the target (fig. 38). Without command, he fired and obtained a target hit.

b. Situation 2. Secondary fire-control equipment, stationary target, HE ammunition.

(1) The gunner immediately re-laid, and noted that the first round fired (fig. 39) struck short of the target and on line with the center of the target.

(2) The gunner mentally noted the point on the sight reticle where the burst appeared and, with the turret controls, moved that point into the center of the target (fig. 40). Without command, he fired the next round and obtained a target hit.

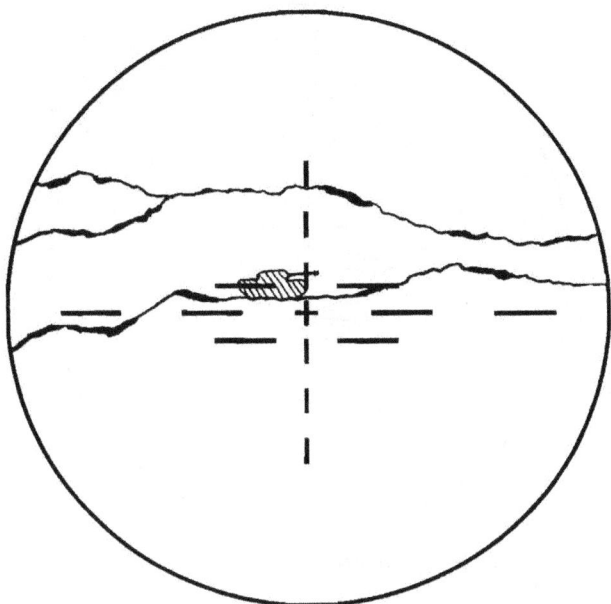

Figure 38. Situation 1. Stationary target.

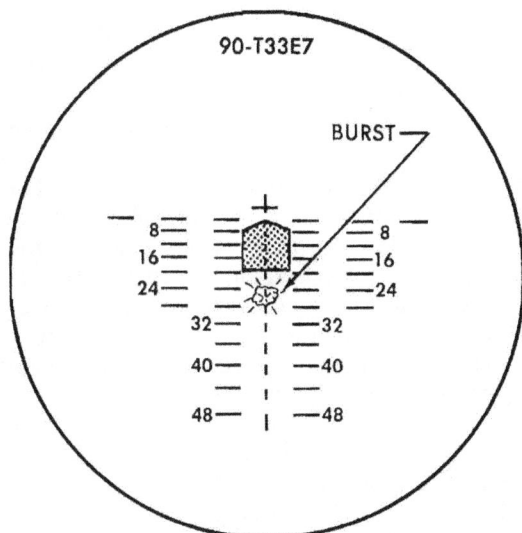

Figure 39. Situation 2. Stationary target.

90-T33E7

8   8
16   16
24   24
32   32
40   40
48   48

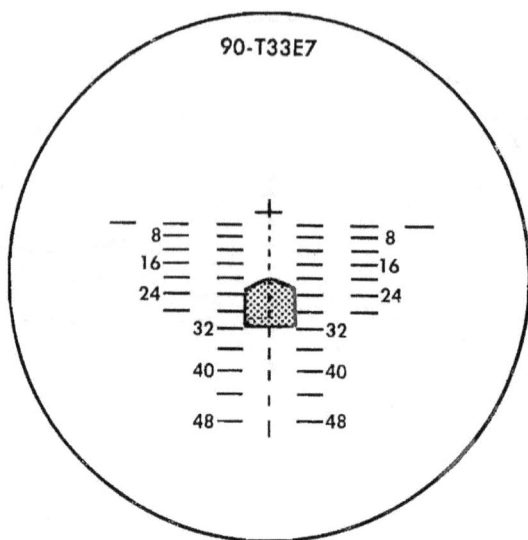

Figure 40. Situation 2. Stationary target.

84. Adjustment of Fire, Alternate Method

The alternate method of adjustment is the tank commander's means of adjustment when the primary method cannot be effectively used due to obscuration, terrain, or extreme range. The alternate method of adjustment involves the use of standard range changes to be announced by the tank commander under certain conditions. The conditions and standard range changes are as follows:

a. When the range finder has been used to determine the initial range to the target and the gunner fails to observe the tracer or burst of his initial round, he will announce LOST. The tank commander then will announce a sensing and subsequent fire command, adding or dropping 200 yards if there was a range error, regardless of range to the target. If the gunner observes this round in his sight, he will apply burst-on-target; however, if this second round also is lost to the gunner, the tank commander will continue with the adjustment, making whatever deflection and range changes he feels are necessary to obtain a target hit. Deflection errors are measured with the binocular, and range changes are made in multiples of 50 yards. If the necessary range change is less than 50 yards, the command may be ADD (DROP) A HAIR.

Note. For practical purposes, a 1-mil elevation change will change the range 100 yards. During the alternate method of adjustment, the gunnner will use the range lines of the gun-laying reticle to make the necessary range change.

b. When the initial range to the target is estimated by the tank commander and the gunner fails to see the tracer or burst, he will announce LOST. The tank commander will then announce a sensing and subsequent fire command, making a range change (if necessary) as follows:

(1) If the estimated range to the target is *1,500 yards or less*, he will add or drop 200 yards of range for the first adjustment. Subsequent adjustments are the same as described in subparagraph *a*, above.

(2) If the estimated range to the target is over *1,500 yards*, he will add or drop 400 yards for the first adjusted round. If the gunner observes this round, he will apply burst-on-target; if not, he will announce LOST, and the tank commander will continue with the adjustment as described in *a*, above.

Note. If an extremely large error is made in the initial *estimated* range, the tank commander may cease fire and announce a new initial fire command.

c. When the gunner, during an adjustment, fails to observe a round *after* applying the burst-on-target method, he will announce LOST. The tank commander will take over and use the alternate method of adustment, making the deflection and range changes he feels are necessary to obtain a target hit.

d. When the gunner and tank commander both fail to observe the round, the gunner will announce LOST. The tank commander will announce LOST and give a subsequent fire command. He may fire another round without changing the range, or he may give a range change to bring the next round to where it can be observed.

e. In all of the above cases, the gunner will apply the announced range change by use of his direct-fire sight and will use the primary method of adjustment whenever possible.

f. The tank commander maintains control of his tank at all times and may take over adjustment of fire at any time. *Once an adjustment of the initial round has been made (by either burst-on-target or subsequent command), or a target hit has been obtained, the standard range rule no longer applies.*

85. Subsequent Fire Commands

a. General. The tank commander issues subsequent fire commands to the gun crew to meet various conditions encountered during firing. These commands are necessary when the gunner has announced LOST, when it is desirous to change the ammunition or fuze, when the tank commander desires to cease fire, and when the tank commander desires to take over the adjustment of fire for any reason. The sequence of elements of the subsequent fire commands used when firing at stationary targets is as follows:

Element	Example
(1) Deflection change (in mils)	RIGHT 3.
(2) Range change (in yards)	ADD 200.
(3) Command to fire	FIRE.

Note. Elements may be omitted from the subsequent fire command if not applicable to an adjustment; therefore, the subsequent fire command may contain one, two, or three elements. If a change of ammunition is desired, it is combined with the command-to-fire element.

b. To Change Ammunition or Fuze. When firing, it may be necessary to designate a different type of ammunition. For example, if a round of shot has penetrated a pillbox or heavy masonry building and the tank commander desires to fire HE through the opening, he commands FIRE HE. This alerts all crewmen to a change in the ammunition. The loader at once loads the HE round, announcing HE—UP to inform the gunner and tank commander that the change has been made. He continues to load HE until he hears CEASE FIRE or another change. The gunner, hearing the change in ammunition, indexes the correct (new) ammunition in the computer. The commander uses the same procedure to change the fuze setting; for example, to change from fuze superquick to fuze delay, he commands FIRE FUZE DELAY. Normally, a chambered round will be fired even though a change in ammunition or fuze has been announced.

Section IV. FIRING AT MOVING TARGETS

86. General

When firing at moving targets, the duties of the gun crew are relatively the same as when engaging stationary targets. A moving target is one which has apparent speed. Targets moving across the line of sight, either horizontally or diagonally, have apparent speed. Targets moving directly toward or directly away from the tank have no apparent speed and are not engaged as moving targets. Proper leading and tracking are of utmost importance when firing at moving targets.

87. Leading

If the gunner fires a round while the gun is aimed directly at a moving target, the target will move out of the path of the projectile, causing the projectile to miss the target. To compensate for this movement of the target, the gunner must aim ahead of the target so that the target and projectile will meet. This technique is called *leading*. To aim ahead of the target, the gunner will use the lead lines in his direct-fire sights. A lead equals 5 mils, regardless of the range to or speed of the target, and is measured from the center of the target. Lead lines in the sight reticle represent 5 mils.

88. Tracking

In order to maintain the proper lead, the gunner must cause the movement of the gun to keep pace with the movement of the target. This manipulation is called *tracking* and is a combination of traversing and changing elevation in order to maintain proper sight alinement. While the tank commander is announcing the first three elements of his initial fire command, he lays the gun for direction, using his power control handle and range finder, and continues to track the target. When the gunner sees the target, he announces IDENTI-FIED, takes control of the turret, and tracks the target at zero lead. The tank commander then ranges on the target with the computer switch in the ON position and, upon completion of ranging, commands ONE LEAD . . . FIRE. The gunner then takes a sight picture of one lead, announces ON THE WAY, and fires. In tracking, the gun should be traversed through and ahead of the target center until the proper lead is applied. The gunner tracks with a smooth, continuous motion, maintaining a constant sight picture before, during, and after firing so that proper sensing and/or adjustments can be made. He will not stop the movement of the gun while he fires; nor will he attempt to ambush the target by moving ahead of it, stopping, and firing when the target reaches the proper lead on the sight reticle.

89. Initial Fire Commands and Firing Duties, Moving Targets

Initial fire commands are the same as those used for engaging stationary targets, with the addition that a lead element will be announced just before the command to fire. Normally, one lead will be used initially, regardless of target speed or range. Listed below are examples of these commands together with the specific duties performed by crewmen in response to each element.

    a. Condition 1.

        (1) Primary fire-control equipment consisting of rangefinder, ballistic computer, and M20 periscope.

        (2) Target: moving tank.

Element	Command	Crewmen's firing duties		
		Tank commander	Gunner	Loader
Alert	GUNNER	Using the range finder and power control handle, lays gun for direction, places ranging reticle above the target, and continues to track the target while giving first three elements of initial fire command. Example: GUNNER, SHOT, TANK.	Turns on turret motor switch if not in power.	Stands by.
Ammunition	SHOT		Using ammunition selector handle, indexes type of ammunition announced in the computer. Turns 90-mm gun switch to ON position.	Selects and loads a round of shot, moves clear of path of recoil, and announces UP. Selects another round of shot.
Range	Omitted.			
Direction	Omitted.			

Target description	TANK	Releases his power control handle when the gunner announces IDENTIFIED; ranges on the target, and upon completion of ranging gives remainder of initial fire command. Example: ONE LEAD, FIRE.	As soon as he sees the target, announces IDENTIFIED, takes control of the turret, and tracks the target at zero lead.	Loads round of shot and continues to load shot without command until CEASE FIRE or a change in ammunition is announced.
Lead	ONE LEAD			
Command to fire	FIRE		Takes proper sight picture, applying one lead. Announces ON THE WAY, and fires.	

b. Condition 2.

(1) Secondary fire-control equipment, consisting of the telescope, rangefinder, and ballistic computer.

(2) Target: moving truck.

Element	Command	Crewman's firing duties		
		Tank commander	Gunner	Loader
Alert	GUNNER	Using range finder and power control handle, lays gun for direction and places ranging reticle above target, while giving first three elements of the initial fire command. Example: GUNNER, SHOT, TANK.	Turns on turret motor switch if not in power.	
AMMUNITION	HE		Using ammunition selector handle, indexes HE in the computer. Turns 90-mm gun switch to ON position.	Selects and loads a round of HE, moves clear of the path of recoil, and announces UP. Selects another round of HE.

110

Range	Omitted	Ranges on the target; after ranging, turns off computer switch and announces the remainder of the initial fire command. Example: ONE LEAD, FIRE.	Using the ammunition selector handle, indexes APT–T33E7. The outer pointer will then indicate the range that the gunner must use on the telescope in order to hit a target with HE ammunition.* *Note.* Either the tank commander or gunner must track the target while the gunner reads the range indicated by outer pointer.	
Direction	Omitted.			
Target description	TRUCK		Announces IDENTIFIED as soon as he sees the target. Then tracks the target at zero lead. Upon hearing the command FIRE, manipulates his controls to place the correct range line and lead on the target, announces ON THE WAY, and fires.	Loads a round of HE and continues to load HE without command until CEASE FIRE or a change in ammunition is announced.

*If the computer is damaged, the gunner may use an aiming data chart to determine the correct range line to use in his telescope.

111

90. Sensing, Moving Target

Rounds fired at moving targets are sensed in relation to the target as when fi..ng at stationary targets (par. 82). Any deflection errors are lead errors, and the actual mil error must be converted to leads and/or fraction of leads.

91. Adjustment of Fire, Moving Target

a. Adjustment of Fire, Primary Method. The burst-on-target method of adjustment for moving targets is the same as for stationary targets, and will be used when the gunner is able to observe the rounds fired. Typical examples of burst-on-target adjustment, using the M20 periscope reticle as well as the telescope reticle, appear below:

    (1) *Situation 1.* Primary fire-control equipment, moving target, shot ammunition.

        (*a*) The gunner immediately re-laid after firing the first round and observed the tracer to pass above and to the right of the target (fig. 41).

        (*b*) The gunner mentally noted the point on the sight reticle where the tracer appeared and, with the turret controls, moved that point to the center of the target (fig. 42). Without further command, he fired and obtained a target hit. Note that he continued to track and increased the lead to compensate for the deflection error.

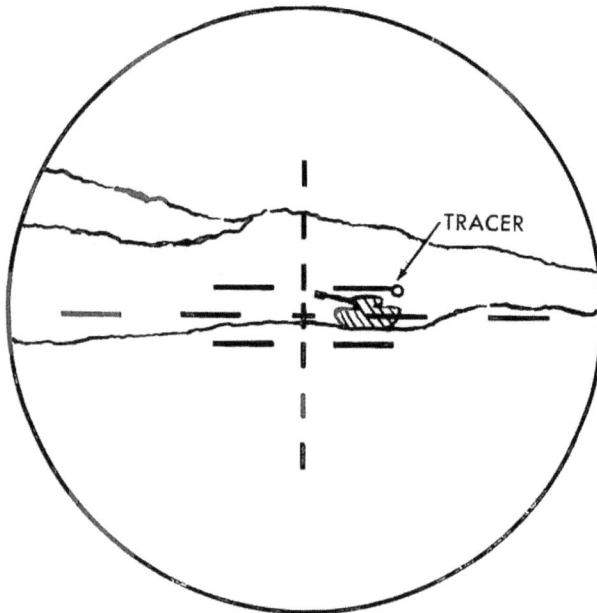

Figure 41. Situation 1. Moving target.

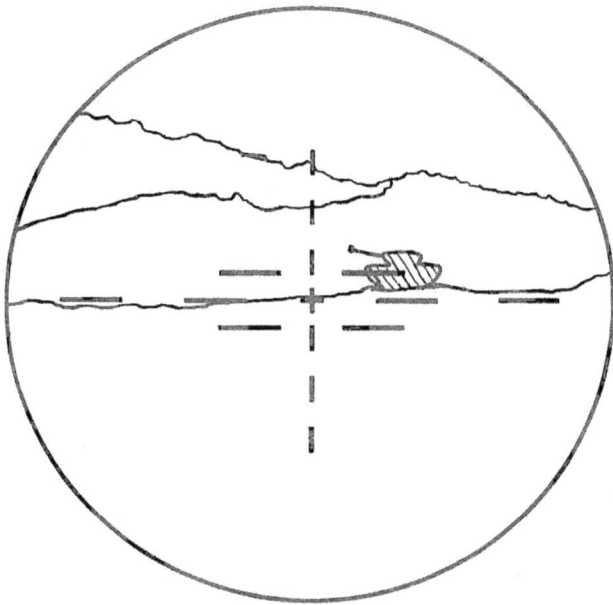

Figure 42. Situation 1. Moving target.

(2) *Situation 2.* Secondary fire-control equipment, moving target, shot ammunition.

  (a) The gunner immediately re-laid and noted that the first round was correct for lead but over in range (fig. 43).

  (b) The gunner mentally noted the point on his reticle where the tracer appeared to pass over the target and, with his turret controls, placed that point of the reticle on the center of the target (fig. 44). He then fired and obtained a target hit.

b. Adjustment of Fire, Alternate Method. The alternate method of adjustment is the tank commander's means of adjusting tank fire when the gunner cannot effectively apply the primary method. The conditions under which this method is applied are the same as those for stationary targets, and a subsequent fire command is given.

92. Subsequent Fire Commands, Moving Target

When necessary, the tank commander will issue subsequent fire commands. Range corrections are announced as prescribed for adjusting on stationary targets (par. 84). Lead corrections are announced as a change in leads rather than in mils. For example, if a round passes behind the center of the target, the tank commander may announce ONE MORE or ONE HALF MORE, and the gunner will increase

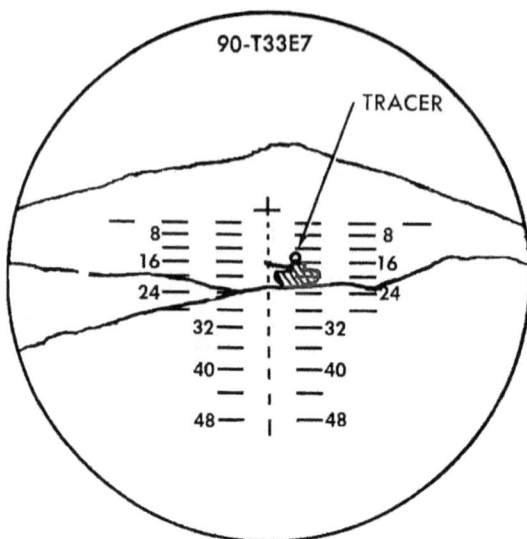

Figure 43. Situation 2. Moving target.

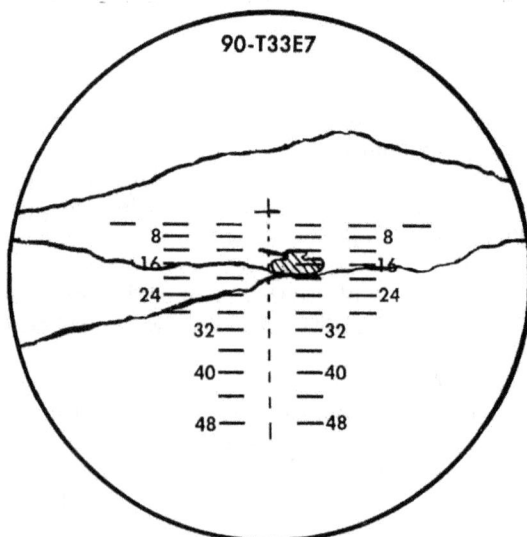

Figure 44. Situation 2. Moving target.

the lead accordingly. Conversely, if the round passes in front of the center of the target, the tank commander may announce ONE LESS or ONE HALF LESS, and the gunner will decrease the lead. The sequence of elements of the subsequent fire command used when adjusting fire on moving targets is as follows:

Element	Example
(1) Range change (in yards)	ADD 200.
(2) Lead change (in leads)	ONE HALF MORE.
(3) Command to fire	FIRE.

Section V. FIRING TANK MACHINEGUNS

93. General

A tank develops its decisive ability when it closes with the enemy in the final assault, and it is in this phase that the tank machineguns play the major role. The machineguns may be fired from moving or stationary tanks, and the targets engaged may be either moving or stationary. These guns furnish a great volume of fire, and a large supply of machinegun ammunition is carried in the tank. The gun crew fires the tank gun only when the machine gun will not accomplish the mission. Therefore, the training of tank crewmen in the methods and techniques of firing the tank machineguns must be emphasized. The tank has two machineguns, the coaxial gun and the turret- or cupola-mounted gun. The technique of firing these machineguns is as follows:

a. Coaxial Machinegun. The commands used to engage either moving or stationary targets with the coaxial machinegun are given in the same sequence of elements as when firing the tank gun (pars. 81 and 89). The coaxial machinegun is fired in bursts of 20 to 25 rounds; the gunner adjusts fire by manipulating the tracer stream into and throughout the target area. An example of an initial fire command follows:

Element	Example
Alert	GUNNER.
Ammunition	CALIBER .30.
Range	FIVE HUNDRED.
Direction	(Omitted).
Description	TRUCK.
Lead	ONE LEAD.
Command to fire	FIRE.

Note. The range is announced mainly for location purposes. The gunner normally fires the machinegun from the battle sight setting. He is indexed on the computer when firing the coaxial machine gun.

b. Turret- or Cupola-Mounted Machinegun. The tank commander fires the turret- or cupola-mounted machinegun at ground or aerial targets. When firing at ground targets, he fires in bursts of 10 to 20 rounds, adjusting fire by manipulating the tracer stream into and in the target area. Against aerial targets, he fires in one continuous burst as long as the target is within range, manipulating the tracer stream into the target by tracking, leading, and observing.

Section VI. SPECIAL SITUATIONS

94. General

Situations arise in combat that prevent the use of normal techniques of fire and which require that substitute means be utilized to insure target destruction. Examples of these conditions are: the destruction of a dangerous surprise target or a rapidly fleeing target, where the battle sight is employed; the destruction of dug-in or defiladed targets by richochet fire; the destruction or neutralization of point and area type targets by use of massed fire; firing during periods of limited visibility with a prepared range card; and firing upon targets from a defiladed position.

95. Dangerous Surprise Targets

Any enemy weapon capable of seriously damaging your tank (for example, a tank or antitank or self-propelled gun), which either has fired at you or is capable of bringing fire to bear upon you immediately, is considered a dangerous surprise target and should be engaged with the battle sight.

96. Fleeing Targets

Any target that is passing rapidly from view to take up a defiladed firing position, or is about to escape destruction, should be engaged with the battle sight. It must be realized that the tank commander will determine the priority in which to take targets under fire and should not necessarily give priority to a target simply by virtue of its movement.

97. Battle Sight

A battle sight is a predetermined range setting combined with a particular type of ammunition. This range and ammunition setting will be determined by the unit commander as the most suitable combination to destroy the immediately dangerous surprise targets which are expected to appear in the area of combat operations. The two elements which compose the unit battle sight will vary according to information concerning the enemy, terrain, and weather. Normally, in a combat operation, the tank gun and the machineguns will be loaded. Unit commanders should include, in their standing operating procedures (SOP), the range and ammunition setting for battle sights. When employing a battle sight, the following fire commands and techniques will be used.

a. Initial Fire Command—Stationary and Moving Targets.

Element	Example
Alert	GUNNER.
Ammunition	} BATTLE SIGHT.
Range	
Description	TANK.
Leads*	ONE LEAD.
Command to fire	FIRE.

*Only when engaging moving targets.

(1) The ammunition element and the procedure for ranging on the target (which are included in the standard initial fire command) will be omitted, since the battle sight includes a predetermined ammunition and range setting.

(2) The loader will continue to load the same type of ammunition until the tank commander changes ammunition or commands CEASE FIRE. In the event a change of ammunition is desired, the tank commander will announce FIRE HE (or any other type of ammunition he desires to fire). The loader, upon hearing the word FIRE and a type of ammunition other than the type previously loaded, will select and continue to load the new type until the tank commander again changes the ammunition element or commands CEASE FIRE.

> *Note:* In the event that a round has been chambered and the ammunition element has been changed, the chambered round should be fired before loading the new type of ammunition.

b. Subsequent Fire Command—Stationary Target. See paragraph 85.

c. Subsequent Fire Command—Moving Target. See paragraph 92.

98. Ricochet Fire

The gunner cannot apply the primary method of adjustment when firing ricochet fire; therefore, the tank commander senses the effect of the fragments upon the ground and utilizes the alternate method of adjustment to bring that effect upon the target. An example of an initial fire command for ricochet fire is as follows:

Element	Example
Alert	GUNNER.
Ammunition	HE DELAY.
Description	TROOPS.
Command to fire	FIRE.

99. Massed Fire

It often is desirable to mass the fire of two or more tanks against certain targets. Controlled concentrated fire by section, platoon, and company produces greater shock effect than does the uncoordinated fire of an equal number of tanks. Regardless of the situation, the target to be engaged by massed fire will be either a point target or an area target. Listed below are examples of the initial fire commands to engage both point and area targets. Fire commands transmitted by radio must be preceded by call signs or other authorized unit designation. For additional information regarding techniques and procedures on massed fire, see FM 17–12.

a. Point Targets. A point target may be described as a lone building, a single antitank gun, a lone tank, or any single target that may be engaged by direct fire. The sequence of the elements is—

Element	Example
Alert	PLATOON.
Ammunition	HE.
Range	MY RANGE 1200.*
Direction	WATCH MY BURST.
Description	RED BRICK BUILDING.
Command to fire	FIRE.

*The range element is announced to assist the other tank commanders in locating the target designated by the burst of the fired round.

b. Area Targets. An area target is a concentration of several targets—for example, a large body of troops, a truck column, a concentration of enemy tanks, etc. The fire command to engage an area target is designed to insure complete target coverage.

(1) To engage a column of tanks from an ambush position, the following fire command would be used:

Element	Example
Alert	PLATOON.
Ammunition	SHOT.
Range	(Omitted).
Direction	DIRECT FRONT.
Description	NO. 2 LEAD TANK.
	NO. 3 SECOND TANK.
	NO. 4 FOURTH TANK.
	NO. 5 LAST TANK.
Command to fire	FIRE.

(2) To engage an area target such as a large concentration of troops, the direction element assigns the sector of fire to the individual tanks. An example is—

Element	Example
Alert	PLATOON.
Ammunition	CALIBER .30.
Range	600.*
Direction	RIGHT FRONT.
	NO. 2 RIGHT FLANK.
	NO. 5 LEFT FLANK.
	NO. 3 RIGHT CENTER.
	NO. 4 LEFT CENTER.
Description	TROOPS.
Command to fire	FIRE.

*The range element is announced only for location purposes. In this case, it designates the center of the sector.

100. Night Firing

To increase the effectiveness of tank weapons during the hours of darkness, it is imperative that commanders plan "on-call" artificial illumination. Artificial illumination is provided by illuminating shells, flares, fires, or searchlights. To deliver effective fire under artificial illumination, an instrument light must be used to illuminate the sight reticle. The rheostat on the instrument light enables the gunner to adjust its brightness. Too much light on the reticle blinds the gunner. For additional techniques on night firing, see FM 17–12.

101. Range Cards

a. A range card is a diagram or sketch of an area, showing the tank's position, the prominent terrain features, and the probable targets, all in their actual relation to positions on the ground. All objects shown on a range card are identified by description, range, quadrant elevation, and deflection (azimuth indicator reading).

b. During combat, range cards are constructed whenever tanks are to be halted for any length of time. The platoon leader normally designates a reference point and the probable targets. Each tank crew then prepares a range card, recording the firing data to all the probable targets visible from its position. The gunner indexes HE on the computer, and the commander then ranges on the reference point. Next, the gunner lays the aiming cross of the M20 periscope on the reference point, zeros the azimuth indicator, and centers the bubble in the elevation quadrant. The range, deflection, and quadrant elevation are recorded. Data for all probable targets and areas in which targets may be expected to appear is obtained as follows:

(1) Under the direction of the tank commander, the gunner indexes the appropriate ammunition on the computer and lays the gun on the probable target, using the manual turret controls.

(2) The tank commander ranges on the target.

(3) Using the manual turret controls, the gunner lays the aiming cross of the M20 periscope on the center of the probable target.

(4) Without disturbing the lay of the gun, the gunner centers the bubble in the elevation quadrant.

(5) The commander then records the range, deflection, and quadrant elevation on the range card.

c. For illustrations and further details on the use of range cards, see FM 17–12.

102. Firing From Defilade

To engage targets from a defiladed position, the tank commander issues a six-element initial fire command in the sequence shown in paragraph 81. The range element is announced in hundreds of yards; and the gunner, referring to a firing table, determines the elevation for the announced range. He applies this to the elevation quadrant, centering the bubble by elevating or depressing the gun. The directional element of the fire command may be given by the reference point method, in which the tank commander lays the gun on a reference point and has the gunner traverse right or left a given number of mils, using the azimuth indicator. The tank commander observes the effect of the fire and uses the alternate method of adjustment to obtain target hits. In firing from defilade, minimum elevation, minimum range, and angle of site must be considered. For details, see FM 17–12.

CHAPTER 5

TANK GUNNERY QUALIFICATION COURSE

Section I. INTRODUCTION

103. General

This chapter contains the prescribed tank gunnery qualification courses and gunnery qualification standards.

104. Purpose and Scope of Courses

a. The purpose of the tank gunnery qualification course is to provide a means of determining the proficiency of the tank crewman in gunnery. The course is designed both to test the gunner and to serve as an adjunct to training in the proper care and use of the weapons and their accessories. The tables of the course will be fired for training purposes before they are fired for record. The standard tank gunnery qualification course covers the gunner's preliminary examination, subcaliber firing exercises, and service firing exercises; the limited course omits the service firing exercises. The tables fired in the two types are—

(1) Standard—tables I through VIII.
(2) Limited—tables I through IV.

b. Each tank crewman must pass the gunner's preliminary examination with a score of 80 percent or higher before firing either the standard or limited course. Gunners firing the standard course must attain a score of 280 or higher on tables I through IV before firing tables V through VIII. Since tank crewmen of the active Army must be considered ready for combat, without further training, after successfully completing the tank gunnery qualification course, a definite classification for such personnel, based on a limited course, must not be made. Therefore, use of the limited course for classification purposes is restricted to tank crewmen of the reserve components. When ranges are not available to active Army personnel for service firing (tables V through VIII), the full limited course, practice and record firing, may be fired twice annually for practice; however, classification of gunners will not be made.

c. The total possible points that can be scored in each phase of qualification are as follows:

Possible points

(1) Gunner's Preliminary Examination _____ 320
(2) Subcaliber firing exercises (tables I–IV) _____ 400
(3) Service firing exercises (tables V–VIII) _____ 400

d. In order to obtain maximum use of facilities, tables may be fired in any order within each group, except that table I will be fired first in the subcaliber exercises, and table V will be fired first in service firing exercises.

105. Classification of Gunners

a. Personnel successfully completing the standard course will be classified as expert gunners, first-class gunners, or second-class gunners.

b. Personnel successfully completing the limited course will be classified, subject to the limitation imposed by paragraph 104*b*, as limited first-class gunner or limited second-class gunner.

c. The required score for each classification is as follows:

Standard course

Classification	Score
Expert gunner	720–800.
First-class gunner	640–719.
Second-class gunner	560–639.
Unqualified	559 or less.

Limited course

Limited first-class gunner	360–400.
Limited second-class gunner	320–359.
Unqualified	319 or less.

106. Ammunition Required

The following table lists the ammunition required to fire the tank gunnery qualification course.

Tables (Fired once for practice and once for record)	Coaxial machinegun			Service HE	Service Shot
	Tracer	Ball	Frangible		
Table I[a]	30				
II[b]	15				
III	30	120			
IV[b] [c]			15		
V[d]					4
VI[d]				4	4
VII[d]					10
VIII[d]				1	
Total practice	75	120	15	5[d]	18
Total record	Same as for practice				

[a] Includes 10 rounds for zeroing and failure to fire single shot.
[b] Includes three rounds for zeroing.
[c] Tracer ammunition may be used if frangible ammunition is not available.
[d] Service firing not included in limited course.

107. Rules for Record Firing

a. Before record firing, the gunner is required to check the condition of the weapon, sights, and ammunition. He is permitted a reasonable length of time to do this.

b. Only the examining personnel, and the necessary personnel of the assigned crew of which the gunner being tested is a member, will be in the tank during record firing. The examiner normally will take the tank commander's post.

c. During firing, the gunner performs all the operations required by the test, without the benefit of coaching or assistance. Other members of the crew perform their normal duties. The assigned crewmen will rotate positions within their tank when firing the tank gunnery qualification course.

d. All exercises will be fired with the sights adjusted by boresighting and zeroing. (The established zero will be used for tables VI through VIII.)

e. If a misfire or other malfunction of the tank gun occurs, the gunner will announce MISFIRE; if the machinegun fails to fire, the gunner will announce STOPPAGE. Thereafter, no one is allowed to touch the gun without the authorization of the examiner. The examining officer notes the time and examines the gun. After the malfunction has been corrected, the crewmen will be permitted to complete the exercise. Unless the malfunction was due to the negligence of the gunner, the time required to correct the malfunction will not be counted against the time allowed the gunner for that particular phase of the test.

f. Prior to record firing, the examining personnel must thoroughly familiarize themselves with their duties and the correct firing procedure.

Section II. GUNNER'S PRELIMINARY EXAMINATION, GENERAL

108. General

The gunner's preliminary examination will be conducted by the company commander and such officer and enlisted assistants as may be necessary. The examination will be given to each member of the tank crew, and a score of 80 percent or more will be required before the crewman is permitted to fire the tables of the tank gunnery qualification course.

109. Table of Possible Scores

GUNNER'S PRELIMINARY EXAMINATION

	Possible points
Materiel tests	
Disassembly of breech mechanism	20
Assembly of breech mechanism	20
Care and maintenance	20
Sight adjustment	50
Putting turret into power operation	10
Testing gunner's quadrant	10
Adjusting elevation quadrant	10
Identification and inspection of ammunition	10
Total possible score	150
Stimulated firing tests	
Direct laying, primary sighting devices	40
Direct laying, secondary sighting devices	30
Use of elevation quadrant	30
Use of gunner's quadrant	30
Adjustment using azimuth indicator and elevation quadrant	40
Total possible score	170
Total possible score, gunner's preliminary examination	320

Note. A gunner must score 256 or more points to be eligible to fire subcaliber or service exercises.

Section III. MATERIEL TESTS, GUNNER'S PRELIMINARY EXAMINATION

110. Test on Disassembly of Breechblock

The breech cover is removed, the gun traveling lock is disengaged, and all devices necessary to remove the breechblock are installed with exception of the breechblock-removing tool. The examiner may assist in the removal of the operating shaft and/or in elevating and depressing the gun as directed by the gunner being tested. The examiner commands, DISASSEMBLE BREECHBLOCK, and starts his timing with a stop watch. The gunner is required to remove and disassemble the breechblock, using the prescribed method. If the disassembly is completed in 3 minutes or less, a credit of 20 points is given. A cut of 5 points is made for every additional 30 seconds or part thereof when more than 3 minutes is taken for this test. To facilitate the handling of parts in disassembly and assembly of the breechblock, all parts should be free from oil and grease. A tarpaulin or shelter half should be placed on the floor of the turret to avoid losing or damaging parts, and the turret lights should be turned on to improve visibility. The examiner will note any particular difficulty encountered by the gunner in removing or replacing any part.

If the difficulty is due to the part being bent, burred, or otherwise damaged, the examiner will note the time required to remove or replace that part and deduct it from the total time required. The closing spring will *not* be removed in this test.

111. Test on Assembly of Breechblock

The examiner commands, ASSEMBLE BREECHBLOCK, and starts the timing with the stop watch. The gunner is required to assemble and replace the breechblock, using the prescribed procedure. If the assembly and replacement is correctly performed in 4 minutes or less, a credit of 20 points is given. A cut of 5 points is made for every additional 30 seconds or part thereof when more than 4 minutes is taken for this test. The examiner may assist in the replacement of the operating shaft and/or in elevating or depressing the gun as directed by the crewman being tested.

112. Test on Care and Maintenance

The examiner commands, PERFORM DAILY LUBRICATION CHECK, INCLUDING CHECKING AND FILLING OF RECOIL SYSTEM. The gunner points out the lubricating points for which he is responsible and, from a display of lubricants and lubricating devices, selects the proper ones to be used for each point. All lubricants will be in the containers in which they are normally issued. He explains the procedure for checking, filling, and bleeding the recoil system. The total possible credit is 20 points. A penalty of 4 points is assessed for each of the following errors:

a. Each point of lubrication missed.

b. Each error in selection of lubricating device or lubricant.

c. Each error in procedure for checking, filling, and bleeding the recoil system.

113. Test on Sight Adjustment

The vehicle is placed in a position where several features or objects, suitable for sight adjustment, are in view. These objects should be at varied known ranges of from 500 to 3,000 yards. The gunner must select a target as near 1,500 yards as possible for boresighting. To conduct the exercise, the examiner places the computer on the phase IV tank out of adjustment in any or all of the following: range setting, ballistic correction, or ammunition setting. He also places out of adjustment the diopter setting and the elevation and deflection boresight knobs on the M20 periscope and on the telescope. The examiner commands, MAKE BORESIGHT ADJUSTMENT. The gunner is required to adjust the computer, the periscope, and the telescope, using the prescribed methods. The total possible credit is 50 points. A score of 40 points is given for placing the computer and the periscope in proper adjustment, and 10 points for proper adjustment of the telescope. No partial credit will be given for sights or fire-control equipment not properly adjusted.

114. Test on Putting Turret Into Power Operation

The examiner commands, PUT TURRET INTO POWER. The gunner is required to put the power traverse mechanism into operation, performing all steps in the prescribed sequence. The total possible credit is 10 points. A penalty of 5 points is assessed for each error.

115. Test on Testing Gunner's Quadrant

The examiner commands, TEST GUNNER'S QUADRANT. The gunner is required to make the end-for-end test on a quadrant which is out of adjustment. The gunner also is required to explain the method of determining the correction. The total possible credit is 10 points. A penalty of 5 points is assessed for each error.

116. Test on Adjusting Elevation Quadrant

The elevation scale and the micrometer scale of the elevation quadrant are placed out of adjustment. The examiner commands, AD-JUST ELEVATION QUADRANT. The gunner is required to adjust both the elevation and the micrometer scales, using the prescribed procedure. No credit is given if the adjustments are not precise. The total possible credit is 10 points.

117. Test on Identification and Inspection of Ammunition

All standard types of ammunition, for both the tank gun and the machineguns, are displayed. Some rounds should have apparent faults, such as a dented cartridge case, badly burred rotating band, or adhering dirt. Lettering and markings on service ammunition are covered. The gunner is required to identify five rounds of ammunition as they are pointed out to him. Color, shape, and size are used as means of identification. The gunner also is required to locate the defects in three rounds that are pointed out to him, and to describe the results of using these rounds without correcting the defect. The total possible credit is 10 points. A penalty of 2 points is assessed for each error.

Section IV. SIMULATED FIRING TESTS, GUNNER'S PRELIMINARY EXAMINATION

118. Test on Direct Laying, Primary Sighting Devices (Computer and M20 Periscope)

In this test, the gunner sets off the correct range and ammunition in the computer and lays the gun for the correct sight picture on four targets.

a. The tank is placed in a position from which several targets can be seen. (The aiming point on the targets should be well defined to eliminate confusion as to correct laying of the gun.) The gunner checks the diopter setting on the M20 periscope, sets the unit battle sight into the computer, and announces READY.

b. The examiner gives a five-element initial fire command while laying the gun for direction with the commander's power control handle and range finder. The examiner starts timing when the ammunition is announced. The gunner indexes the announced ammunition, makes final lay on the aiming point of the indicated target, and announces ON THE WAY. The examiner checks the initial sight picture and the computer to determine whether the correct range and ammunition have been indexed. If the gunner performs this exercise correctly within 10 seconds, he receives 10 points. He will receive no credit if he fails to complete the exercise correctly within 10 seconds.

c. The test then is repeated three times, using the procedure outlined in *b* above, for a total of four trials.

d. The total possible score for this test is 40 points.

119. Test on Direct Laying, Secondary Sighting Devices

In this test, the gunner must lay the gun correctly three times, using the telescope and the computer or an aiming data chart.

Note. If the ammunition element is other than that for which the sight is graduated, the computer or an aiming data chart must be used. For an example of an initial fire command for this exercise, together with an explanation of how the gunner uses the computer, see paragraph 81c. For an explanation of the use of an aiming data chart, see FM 17–12.

a. The gunner adjusts the diopter setting and announces READY.

b. The examiner gives a five-element initial fire command while laying the gun for direction with the commander's power control handle and range finder. Timing starts and the test begins when the ammunition element is announced. The gunner refers to the computer or an aiming data chart, makes final lay of the gun on the target, and announces ON THE WAY. The examiner checks the initial sight picture, using the gunner's telescope, and checks the computer or aiming data chart to determine whether the gunner selected the correct data for the range and ammunition specified. If the gunner performs this exercise correctly within 10 seconds, he receives 10 points. He will receive no credit if he fails to complete the exercise correctly within 10 seconds.

c. The examiner then gives a subsequent fire command, using a deflection change of not more than 10 mils and a range change of not more than 400 yards, as prescribed in the alternate method of adjustment. The gunner lays the gun with the corrected sight picture and announces ON THE WAY. The examiner checks the sight picture, using the gunner's telescope. If the gunner performs this exercise correctly within 5 seconds, he receives 10 points. He will receive no credit unless he has obtained the correct sight picture within 5 seconds.

d. The examiner then gives a second subsequent command, using the procedure outlined in *c* above.

e. The total possible score for this test is 30 points.

120. Test on Use of Elevation Quadrant

In this test, the gunner lays the gun for range, using the elevation quadrant.

a. The examiner announces a range in yards. Using the computer as an aiming data chart, the gunner determines the elevation corresponding to the announced range, applies this elevation to the elevation quadrant, and centers the bubble, using the manual elevation control handle. He then announces ON THE WAY. The examiner checks the elevation setting and the lay of the gun.

b. The examiner then announces two subsequent ranges, and the same procedure is followed.

c. No credit will be given if an improper quadrant setting is used or if the bubble is not accurately centered. For each trial correctly performed in 9 seconds or less, the gunner receives 10 points. He will receive no credit if the time exceeds 9 seconds. The total possible score for this test is 30 points.

121. Test on Use of Gunner's Quadrant, M1

In this test, the gunner is required to set an elevation on the M1 quadrant and to place the quadrant properly on the quadrant seats.

a. The gunner sets the quadrant at zero. The examiner announces an elevation. The gunner applies this elevation, properly seats the quadrant on the breech, and announces SET. The examiner checks the elevation and the position of the quadrant.

b. The examiner then announces two subsequent elevations, one of which will be to an even tenth of a mil, and the same procedure is followed.

c. No credit will be given if the elevation is incorrect or if the quadrant is seated improperly. For each trial correctly performed in 7 seconds or less, the gunner receives 10 points. He will receive no credit if the time exceeds 7 seconds. The total possible score for this test is 30 points.

122. Test on Adjustment Using Azimuth Indicator and Elevation Quadrant

a. The examiner sets the azimuth indicator at zero and the elevation quadrant at an arbitrary elevation, centering the bubble with the turret controls. This takes the place of laying the gun from an initial fire command.

b. The examiner then gives four subsequent commands. The deflection shifts should be in multiples of 5 mils and should not exceed 100 mils. The gunner follows the commands, using the azimuth indicator for deflection changes and the elevation quadrant for range changes. After each azimuth indicator and quadrant setting has been applied and the bubble centered, the gunner announces ON THE WAY. The examiner checks the azimuth indicator and quadrant setting and the bubble of the elevation quadrant.

c. No credit will be given if any setting is incorrect or if the bubble of the quadrant is not centered. For each trial correctly performed in 20 seconds or less, the gunner receives 10 points. He will receive no credit if the time exceeds 20 seconds. The total possible score for this test is 40 points.

Section V. SUBCALIBER FIRING EXERCISES

123. General

All subcaliber firing will be conducted with the coaxial machinegun. When single shots are fired, ammunition must be loaded with alternate dummy rounds, or a single-shot device may be used. Controller, Single-shot, Caliber .30-Caliber .50, or CONARC-approved training aid No. 1, single-shot device, may be used. Targets will be physically scored during all record firing.

124. Table I—Subcaliber Manipulation Exercise, 1,000-Inch

a. The purpose of this exercise is to test the gunner's ability to 'manipulate the turret controls rapidly and to engage stationary targets firing single-shot. Figure 45 illustrates the target layout. The 4 x 4-inch targets may be stapled or tacked onto staves.

Note: The coaxial machinegun is zeroed for 1,000-inch firing using the telescope sight.

b. The gunner lays on the left (No. 5) target. At the command, COMMENCE FIRING, he lays the aiming point of the sight reticle on the center (No. 1) target and fires one round. He then fires one round at each of the remaining four targets in the order in which

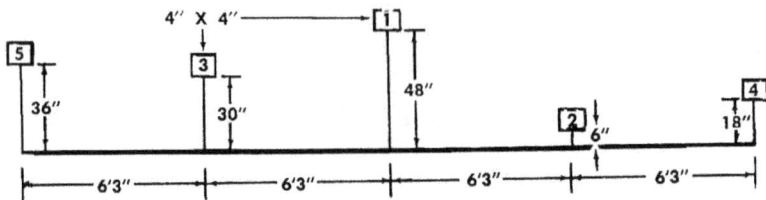

Figure 45. Manipulation target layout.

they are numbered. Time is recorded from the command, COM-
MENCE FIRING. At the end of the allowed time, the examiner
will command, CEASE FIRING. Rounds fired after this command will be scored as misses.

c. The exercise consists of four trials as outlined in *b* above, two in
manual traverse and two in power traverse. Credit will be given in
accordance with table I, below.

d. The exercise will be scored as follows:

Table I

Possible score: 100

Trials	Number rounds	Time (seconds)	Points each hit
1st—Manual traverse_____	5	35	5
2d—Manual traverse_____	5	35	5
3d—Power traverse_____	5	35	5
4th—Power traverse_____	5	35	5

Note. Targets will be scored and marked, and replaced when required.

e. See figure 46 for illustration of an appropriate score card.

125. Table II—Subcaliber Shot Adjustment, Moving Target Exercise, 200 Feet

a. The purpose of this exercise is to test the gunner's ability to lead,
track, fire on, and adjust fire on moving targets prior to firing service
ammunition.

b. In this exercise, the gunner is required to fire tracer (single-shot)
at moving targets on a 200-foot range, using the coaxial machinegun.
The target speed is controlled and the sight is zeroed so that when
the gunner takes the correct sight picture of one lead with the correct
range applied, he will obtain a target hit.

c. Targets are mounted on a 6 x 6-foot panel as shown in figure 47.
The target shown is adequate to fire three tanks simultaneously; if it
is desired to fire more tanks, targets may be pulled in tandem. The
speed of the targets should be approximately 5 miles per hour so that

Co __A__

Bn __1ST__

NAME *SMITH JOHN J.*

RANK *PFC* SN *38132977*

DATE __1 FEB 54__

75

Total Score

TANK GUNNERY QUALIFICATION COURSE SCORE CARD

100 points
Possible

TABLE I (SUBCALIBER MANIPULATION EXERCISE—1000-INCH)

TRIALS	NUMBER OF ROUNDS	POSSIBLE	MAX TIME (SEC)	TARGET HITS					SCORE
				1st RD	2d RD	3d RD	4th RD	5th RD	
1st—Manual	5	25	35	5	5	5	0	0	15
2d—Manual	5	25	35	5	5	0	5	5	20
3d—Power	5	25	35	5	5	5	5	0	20
4th—Power	5	25	35	5	5	0	5	5	20
								TOTAL SCORE	75

A. Five points for each target hit. No credit for hits obtained after time limit.

B. Add score column to obtain total score for this exercise.

Satisfactory Score _____ 70 points.

Lt Leonidas R. James

Examining Officer's Signature

Figure 46. Score card for table I.

the gunner can track the target on the manual trials. The machine-gun will be zeroed to hit the center of the target when one lead is taken.

d. When a moving target range is not available, target tanks may be utilized for this exercise if available and desirable. Ranges for target tanks should be about 400 yards.

e. Using the M20 periscope, with the unit battle sight indexed on the computer, the coaxial machinegun is zeroed to hit the center of the target at a distance of 200 feet when one lead is taken. If the primary sight will not converge with the coaxial machinegun at a distance of 200 feet, the secondary sight may be used and zeroed to hit the center of the target with a sight picture of 800 yards when one lead is taken; however, every effort should be made to zero the primary sight. A target traveling 5 miles per hour requires only one half of a lead.

131

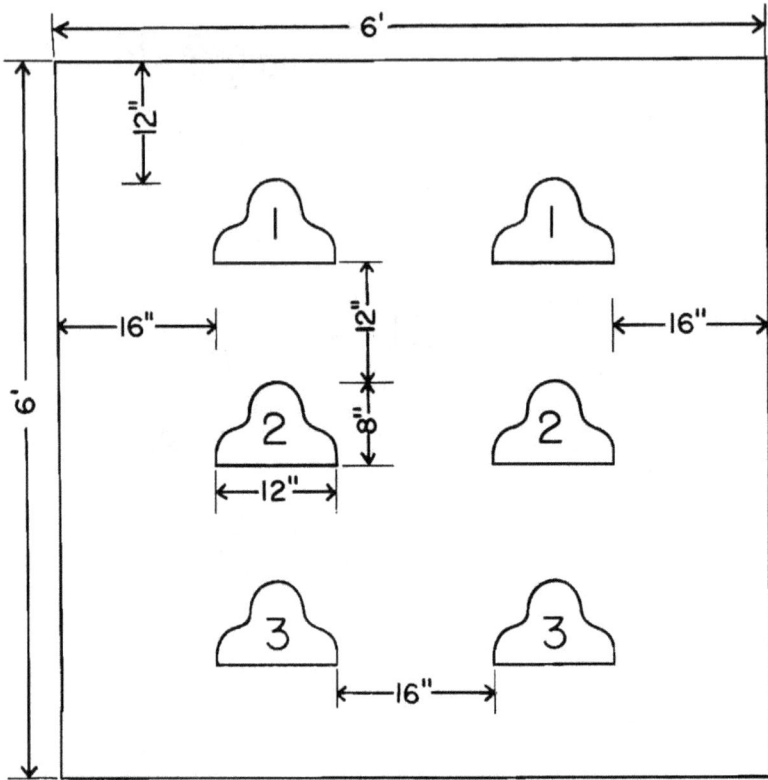

Figure 47. Moving target layout.

The gun, therefore, must be zeroed on a stationary target with one half lead in order for the gunner being tested to hit the moving target with a sight picture of one lead. Figure 48 shows the proper zero for both the primary and secondary sights to fire on a target traveling from left to right.

    f. The exercise is conducted as follows:

        (1) As the targets move along the course, the examiner gives a fire command designating the target by number. The gunner then fires three rounds at the designated target. The examiner records time from the command FIRE until after the third round is fired.

        (2) The exercise consists of four trials as outlined in (1) above; two trials are fired at the lead target in manual traverse and two at the rear target in power traverse.

    g. The exercise will be scored as follows:

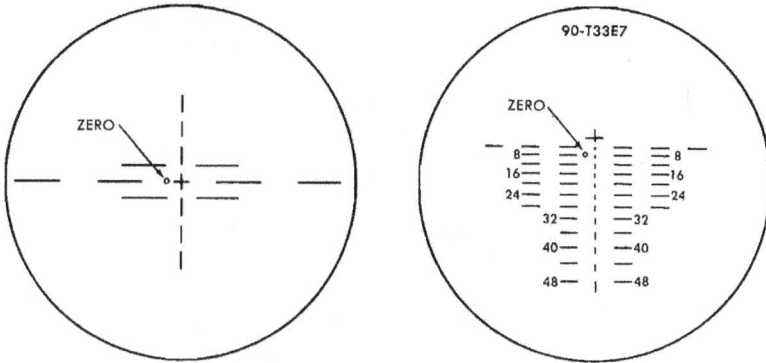

Figure 48. Zeroing for table II.

Table II

Possible score: 100

Trials	Number rounds	Possible	Minus	Score
Manual_____	3	25	_____	_____
Manual_____	3	25	_____	_____
Power_____	3	25	_____	_____
Power_____	3	25	_____	_____

Cuts for each target:

  Failure to fire first round within 5 seconds_____ 5 points.
  Failure to complete trial in 15 seconds_____ 5 points.
  Each round that fails to hit target_____ 5 points.

h. See figure 49 for illustration of an appropriate score card.

126. Table III—Moving Tank Exercise (Stationary Target)

a. The purpose of this exercise is to test the gunner's ability to fire the coaxial machinegun from a moving tank at stationary targets.

b. In this exercise, the gunner fires 150 rounds from the coaxial machinegun at targets representing infantry.

c. The length of the course (fig. 50) will be approximately 800 yards. The end of the course will be marked by a white flag on each side of the runway. Targets will be kneeling-type (E) silhouettes. Five groups of four silhouettes will be placed not more than 10 yards nor less than 5 yards from the sides of the tank runway, alternately on the left and right sides. One of these groups will be placed at each of the

133

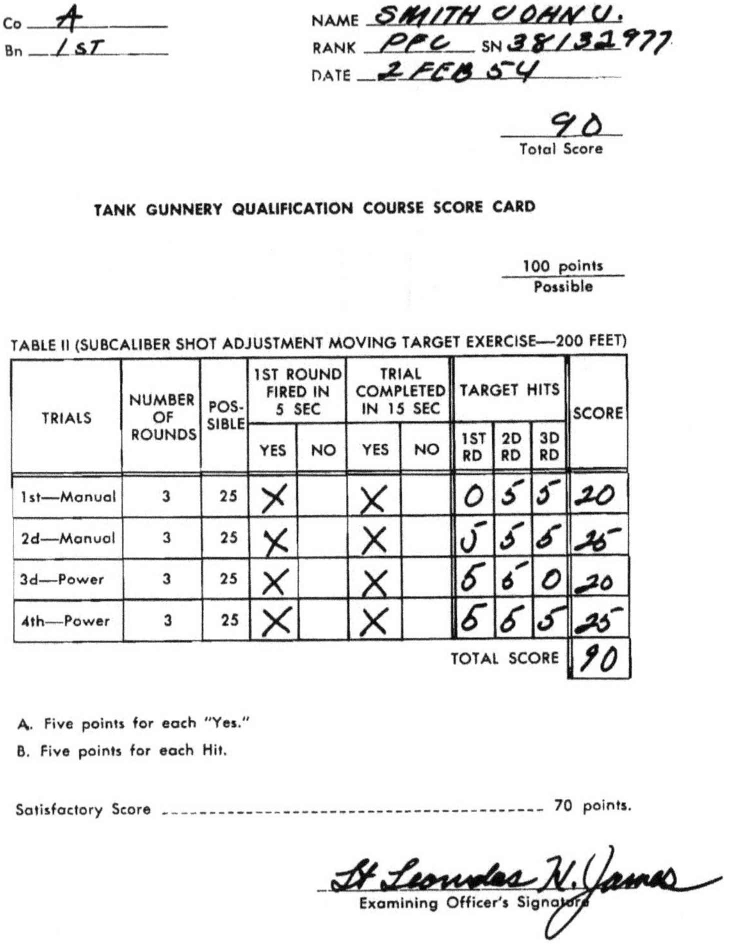

Co __A__

Bn __1ST__

NAME _SMITH JOHN J._

RANK _PFC_ SN _38132977_

DATE _2 FEB 54_

$$\frac{90}{\text{Total Score}}$$

TANK GUNNERY QUALIFICATION COURSE SCORE CARD

$$\frac{\text{100 points}}{\text{Possible}}$$

TABLE II (SUBCALIBER SHOT ADJUSTMENT MOVING TARGET EXERCISE—200 FEET)

TRIALS	NUMBER OF ROUNDS	POS-SIBLE	1ST ROUND FIRED IN 5 SEC		TRIAL COMPLETED IN 15 SEC		TARGET HITS			SCORE
			YES	NO	YES	NO	1ST RD	2D RD	3D RD	
1st—Manual	3	25	X		X		0	5	5	20
2d—Manual	3	25	X		X		5	5	5	25
3d—Power	3	25	X		X		5	5	0	20
4th—Power	3	25	X		X		5	5	5	25
								TOTAL SCORE		90

A. Five points for each "Yes."

B. Five points for each Hit.

Satisfactory Score _____ 70 points.

Lt Leonidas N. James
Examining Officer's Signature

Figure 49. Score card for table II.

following ranges from the starting point: 200 yards, 350 yards, 450 yards, 550 yards, and 700 yards. A sixth group of five silhouettes will be placed 200 yards beyond the end of the course and in direct line with the center of the runway. Each group of silhouettes should cover an area 4 yards wide and 4 yards deep. A red flag will be placed on the edge of the runway, 50 yards from each silhouette group in the direction of the starting line.

d. The terrain selected for the course will be such that the tank can maintain an average speed of 5 miles per hour.

e. The ammunition will be loaded four ball to one tracer.

f. The test is conducted as follows:

(1) The gunner will make only one run over the course while

being tested. Fire will cease on a target group when the tank reaches the red flag 50 yards from that group. All hatches will be closed while the tank is between the starting line and the white flags.

(2) The tank will not stop until it reaches the white flags at the end of the runway. At this point, firing will cease and the gun will be cleared. The tank will be required to maintain an average speed of 5 miles per hour between the starting line and the white flags.

(3) Separate courses may be set up and used concurrently, provided they are at least 30 yards apart.

(4) The examiner should follow behind the tank in a vehicle and should control the exercise by radio. Assistants will follow the tank to mark targets and score.

(5) An assistant will check the number of rounds in each belt before the tank begins the run.

g. The exercise will be scored as follows:

Table III

Possible score: 100

Target groups	Number rounds	Possible	Minus	Score
1	25	16 points	------	------
2	25	16 points	------	------
3	25	16 points	------	------
4	25	16 points	------	------
5	25	16 points	------	------
6	25	20 points	------	------

Total ----- 100 points

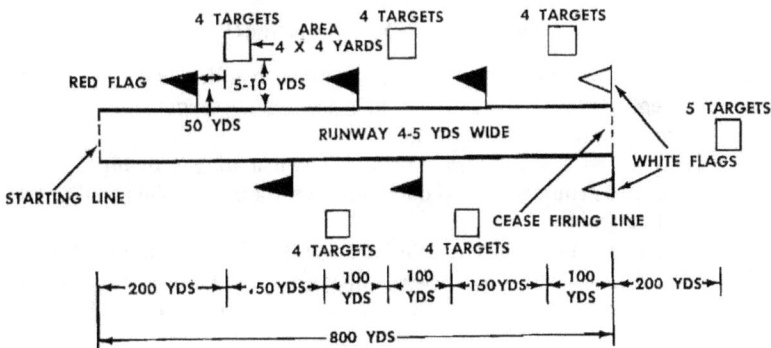

Figure 50. Range setup—moving tank exercise.

NAME *SMITH JOHN J.*
RANK *PFC* SN *38132977*
DATE __*3 FEB 54*__

__*80*__
Total Score

TANK GUNNERY QUALIFICATION COURSE SCORE CARD

100 points
Possible

TABLE III (MOVING TANK EXERCISE, STATIONARY TARGETS)

TARGET GROUPS	NUMBER OF ROUNDS	POSSIBLE	TARGET HITS					SCORE
			NR 1	NR 2	NR 3	NR 4	NR 5	
1	25	16	4	4	0	4		12
2	25	16	4	0	4	4	*(Applies to target group Nr 6 only)*	12
3	25	16	4	4	4	4		16
4	25	16	0	4	4	4		12
5	25	16	4	4	4	0		12
6	25	20	4	4	4	0	4	16
							TOTAL SCORE	80

A. Four points for each target hit. (No credit for target hits after passing red flag of each group.)

B. No credit if tank does not maintain a speed of 5 mph.

Satisfactory Score _____ 70 points.

Lt Leonidas K. James
Examining Officer's Signature

Figure 51. Score card for table III.

(1) Four points will be awarded for each silhouette that is hit regardless of the number of hits in each target. The maximum score is 100 points.

(2) No credit will be given for hits on a target group if that target group was fired on after the tank passed the red flag for that group.

(3) The gunner receives no credit for the course if his tank fails to maintain a 5-mile-per-hour speed, but he will be retested.

h. See figure 51 for illustration of an appropriate score card.

127. Table IV—Auxiliary Fire Control Exercise

a. Purpose. The purpose of this exercise is to test the ability of the gunner in the proper use of the tank's auxiliary fire-control instruments.

b. Target Layout.

(1) The impact area selected should be fairly flat, and the surface should be of dirt or sand.

(2) An aiming post, to serve as a gun reference point and 1,000-yard range marker, is placed midway within the width of the individual tank's impact area at a distance of 117 feet from the muzzle of the coaxial machinegun.

(3) At varying simulated ranges (500–1,500 yards), small vertical targets (4 x 6-inch cards) will be fastened to the ground. The bottom edges of the targets will be folded to provide a fastening surface and a 4 x 4-inch target facing the tank. See figure 52 for target layout.

c. Preparation by Examiner.

(1) With the coaxial machinegun properly mounted and adjusted, the examiner fires until he obtains a hit on the base of the gun reference point.

Note. A 4 x 4-inch target is placed immediately behind the aiming post as an aid in zeroing.

(2) Without disturbing the lay of the gun, the examiner will—

(*a*) Center the bubble of the elevation quadrant with the micrometer knob.

(*b*) Loosen the micrometer knob and slip the scales until a reading of 50 mils appears opposite the index, then tighten the micrometer knob.

(*c*) Place a fine chalk or pencil line on the elevation scales opposite the index. This is to avoid 100-mil errors when the gun is elevated between problems.

(*d*) Zero the azimuth indicator.

(*e*) Adjust the boresight point of the telescopic sight (using the boresight knobs) to the point of strike on the gun reference point.

(3) After zeroing on the reference point, the examiner will determine the azimuth indicator reading and elevation quadrant reading to the center of each target in the impact area and will record this data on a reference card (fig. 53). The elevations are converted to ranges, which also are recorded on the reference card.

Note 1. Make certain it is clearly understood that this is a REFERENCE card and *not* a RANGE card.

Note 2. Elevations are converted to ranges by reference to the following: The elevation for a range of 1,000 yards equals 50 mils. A 100-yard range change is effected by making a 1-mil change in elevation.

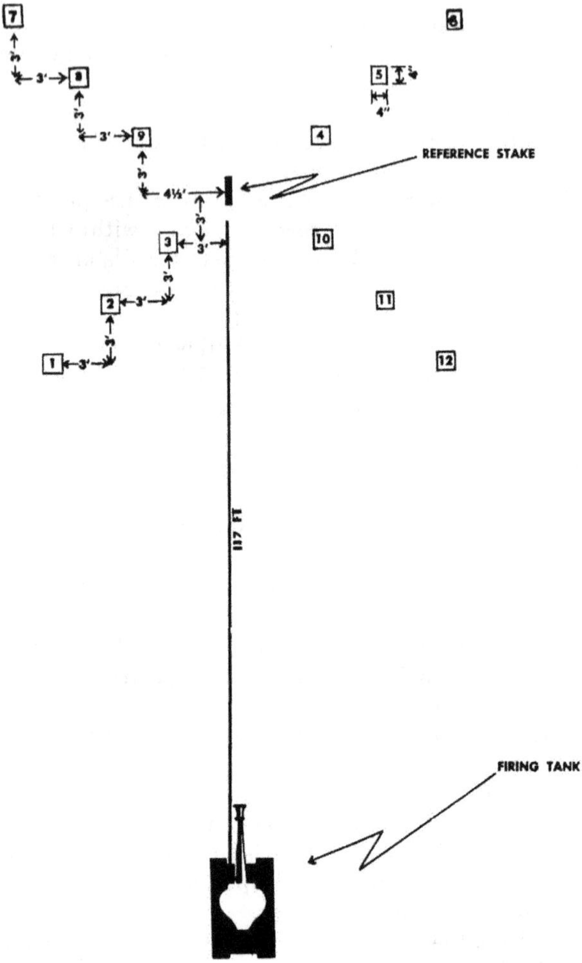

Figure 52. Target layout, table IV.

(4) The gunner will use a 1-mil change in elevation for each 100-yard range change desired.

d. Procedure for Testing.

(1) The gunner will be required to engage four targets, using the auxiliary fire-control instruments. The direct-fire sights will be covered during the exercise. Three rounds of caliber .30 frangible ammunition will be fired at each target; tracer ammunition may be used if frangible ammunition is not available. (The single-shot device can be used, or dummy rounds can be inserted between live rounds to permit firing single shots.)

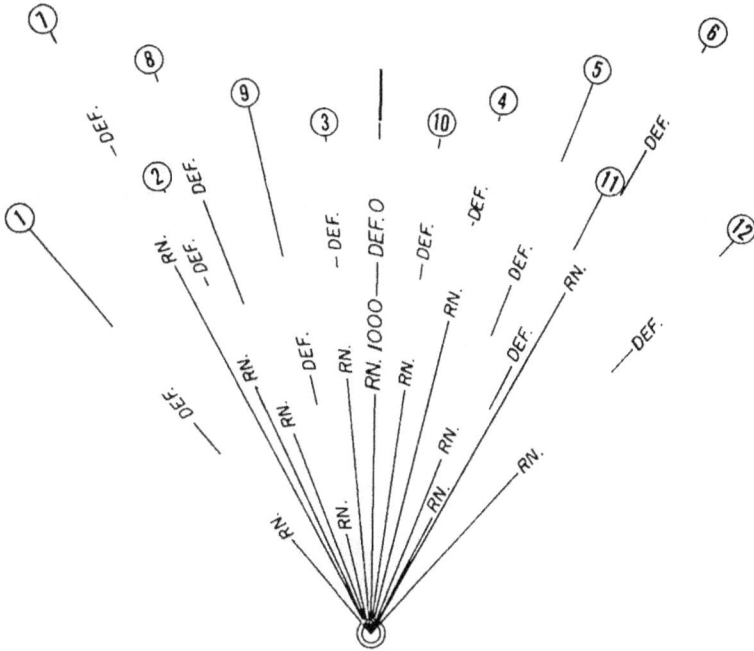

Figure 53. Reference card for table IV.

(2) The examiner, acting as tank commander, will issue correct initial fire commands, sense each round, and give the necessary subsequent commands for each target. For example—

Element	Example
Alert	GUNNER.
Ammunition	HE.
Range	1500.[a]
Direction	FROM REFERENCE POINT, RIGHT 30.[b]
Description	ANTITANK.
Command to fire	FIRE.[c]

[a] The announced range will be the correct range to the target.
[b] Deflection to the target will be based on previously computed data.
[c] Time for the exercise commands will be recorded from this element.

Subsequent commands will be in accordance with the alternate method of adjustment. Time for subsequent rounds also will be recorded from the command FIRE.

(3) The gunner, in executing these commands, will use the specially prepared firing table and the elevation quadrant for initial and subsequent ranges.

(4) In making initial and subsequent deflection shifts, the gunner will use the azimuth indicator. At the conclusion of firing on each target, the gun is returned to the reference point

and the bubble in the elevation quadrant is centered with a reading of 50 mils.

e. Scoring. The exercises will be scored as follows:

Table IV

Possible score: 100

Trial	Number rounds	Possible points	Minus	Score
Target 1_____	3	25	_____	_____
Target 2_____	3	25	_____	_____
Target 3_____	3	25	_____	_____
Target 4_____	3	25	_____	_____

Cuts for each target:
    Failure to get first round off in 20 seconds_____ 5 points.
    Failure to get second round off in 10 seconds_____ 5 points.
    Failure to get third round off in 10 seconds_____ 5 points.
    Each target miss over one_____ 5 points.

f. Score Card. See figure 54 for illustration of an appropriate score card.

Co __A__

Bn __1st__

NAME __SMITH, JOHN J.__

RANK __PFC__ SN __38132977__

DATE __3 FEB 54__

__80__
Total Score

TANK GUNNERY QUALIFICATION COURSE SCORE CARD

__100 points__
Possible

TABLE IV (AUXILIARY FIRE CONTROL EXERCISE)

TRIALS	Nr OF RDS	POS-SIBLE	1ST RD IN 20 SEC		2D RD IN 10 SEC		3D RD IN 10 SEC		TARGET HITS			CUTS	SCORE
			YES	NO	YES	NO	YES	NO	1ST RD	2D RD	3D RD		
Target 1	3	25		X	X		X		O	X	X	5	20
Target 2	3	25	X		X		X		O	O	X	5	20
Target 3	3	25	X		X			X	X	O	O	10	15
Target 4	3	25	X		X		X		X	X	O	O	25
											TOTAL SCORE		80

A. CUTS: Cut five points for each "NO."

B. Cut five points for each target miss over one (maximum 10 points).

Satisfactory Score --- 70 points.

Lt Lemidas K. Jones
Examining Officer's Signature

Figure 54. Score card for table IV.

Section VI. SERVICE FIRING EXERCISES

128. General

The service ammunition tables are fired only after the gunner has qualified in tables I through IV. Ammunition for both practice and record firing must be of the same lot. Targets will be accurately scored during a record firing; BC scopes or similar viewing instruments may be used for this purpose.

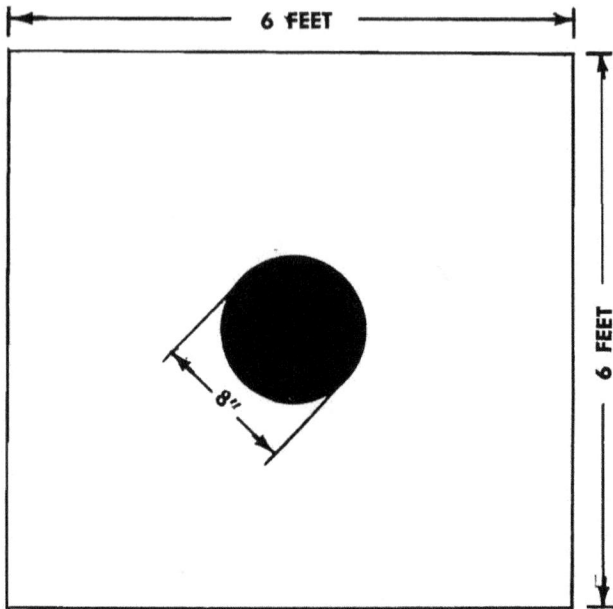

Figure 55. Target panel for zeroing exercise.

129. Table V—Zeroing Exercise, Service Firing

This exercise is designed to test the gunner's ability to zero the M20 periscope and the telescope, using shot ammunition on a 6 x 6-foot panel target at a range as near 1,500 yards as possible (fig. 55).

a. The gunner boresights the M20 periscope and the telescope on the zeroing target and applies the emergency zero.

b. He indexes the ammunition and range on the computer, lays the aiming cross on a definite aiming point in the center of the zeroing target, and fires three rounds, re-laying carefully after each round.

c. Without disturbing the lay of the gun on the aiming point, the gunner next adjusts the aiming cross of the periscope and the appropriate range line of the telescope to the center of the shot group by means of the boresight knobs.

d. After relocking the boresight knobs, he re-lays on the definite aiming point in the center of the zeroing target and fires one check round to verify the zero.

e. There is no time limit for this exercise. The exercise will be scored as follows:

Note. The examiner will boresight and zero the range finder.

142

No.	Item	Possible	Minus	Score
1	Correct boresighting procedure, M20 periscope and telescope, total value.	-----------	30 points___	
	For improper selection of boresight point, cut.	6 points____	-----------	
	For failure to remove all superelevation prior to boresighting, cut.	6 points____	-----------	
	For inaccuracy in laying aiming crosses of reticles on boresighting point observed through bore, cut.	6 points____	-----------	
	For failure to lock boresight knobs after placing aiming crosses on boresight point, cut.	6 points____	-----------	
	For failure to set slip scales properly on boresight knobs on M20 periscope, cut.	6 points____	-----------	
2	Correct zeroing procedure, M20 periscope and telescope, total value.	-----------	30 points___	
	For failure to index proper emergency zero, cut.	6 points____	-----------	
	For failure to index proper ammunition on computer, cut.	6 points____	-----------	
	For failure to index proper range on computer, cut.	6 points____	-----------	
	For failure to place proper range line of telescope on center of shot group, cut.	6 points____	-----------	
	For failure to use boresight knobs in manipulating aiming point of reticles on center of shot group, cut.	6 points____	-----------	
3	Accuracy of zero, total value_____	-----------	40 points___	
	If check round strikes: Within 14 inches from center of aiming point, no cut.			
	More than 14 and less than 18 inches from center of aiming point, cut.	10 points___	-----------	
	More than 18 and less than 24 inches from center of aiming point, cut.	20 points___	-----------	
	More than 24 inches from center of aiming point, cut.	40 points___	-----------	

Note: A physical check of the target will be made to insure positive scoring of the check round.

Co A

Bn 1ST

NAME SMITH, JOHN J.

RANK PFC SN 38132877

DATE 4 FEB 54

84

Total Score

TANK GUNNERY QUALIFICATION COURSE SCORE CARD

100 points

Possible

TABLE V (ZEROING EXERCISE FOR M48 TANKS)

NR 1	BORESIGHTING	YES	NO	SCORE
(30 points) 6 points for each YES.	1. Properly selected boresight point.	X		6
	2. Removed all superelevation prior to boresighting.	X		6
	3. Properly laid aiming cross on boresight point.	X		6
	4. Boresight locking levers locked after placing aiming cross on boresight point.	X		6
	5. Slip scales set properly on boresight knobs of range finder and periscope.	X		6
NR 2	ZEROING PROCEDURE			
(30 points) 6 points for each YES.	1. Indexed proper emergency zero.	X		6
	2. Indexed proper ammunition on ammunition scale.		X	0
	3. Indexed proper range on range scale.	X		6
	4. Placed proper range on telescope or periscope.	X		6
	5. Used boresight knobs in manipulating aiming point of reticles on center of shot group.	X		6
NR 3	ACCURACY OF ZERO			
40 points for YES.	Check round within 14 inches from the center of aiming point.		X	0
30 points for YES.	Check round more than 14 and less than 18 inches from center of aiming point.	X		30
20 points for YES.	Check round more than 18 and less than 24 inches from center of aiming point.		X	0
			TOTAL SCORE	84

Satisfactory Score -- 70 Points

Lt Leonidas K. James

Examining Officer's Signature

Figure 56. Score card for table V.

f. See figure 56 for illustration of an appropriate score card.

130. Table VI—Service Firing Exercise, Stationary Targets at Variable Ranges (Shot and HE Adjustment)

a. This exercise is designed to test the gunner's ability to utilize the primary sighting equipment and the burst-on-target method of adjustment while firing service ammunition at stationary targets. The exercise is more realistic if tanks move between problems. The gunner will fire at four separate targets (two shot and two HE). The true range to the target will be fed into the computer automatically, and the gunner will apply the burst-on-target method of adjustment for subsequent rounds if necessary. The gunner will be limited to two rounds of ammunition for each of the four targets he is to engage.

b. The examiner will indicate each of the targets by issuing an initial fire command. The examiner will lay the gun for direction, using the commander's power control handle and the rangefinder. Time for each problem starts when the command FIRE is announced in the initial fire command.

c. If the first round is not a target hit, the gunner will use the burst-on-target method of adjustment to fire the subsequent round. If the target is hit on the first round, full credit will be given and the second round *will not be fired.*

d. Targets to be used for this exercise are 3 x 5-foot cloth panels for targets 1 and 2, and 6 x 6-foot cloth panels for targets 3 and 4. One HE and one shot problem will be fired at each size target.

e. The exercise will be scored as follows:

Table VI

Possible score: 100

Trial	Range	Number rounds	Possible	Minus	Score
Target 1 _ _ _ _ _	800–1,100 yards _ _ _ _ _ _ _ _ _	2	25	_ _ _ _ _ _ _ _	_ _ _ _ _ _ _ _
Target 2 _ _ _ _ _	1,100–1,500 yards _ _ _ _ _ _ _	2	25	_ _ _ _ _ _ _ _	_ _ _ _ _ _ _ _
Target 3 _ _ _ _ _	1,500–1,800 yards _ _ _ _ _ _ _	2	25	_ _ _ _ _ _ _ _	_ _ _ _ _ _ _ _
Target 4 _ _ _ _ _	1,800–2,000 yards _ _ _ _ _ _ _	2	25	_ _ _ _ _ _ _ _	_ _ _ _ _ _ _ _

Cuts for each target:
    Failure to fire first round within 15 seconds _ 5 points.
    Deduct 1 point for each additional second over 15 required to fire   5 points.
        fire round, up to 20 seconds.
    Failure to hit target with first round _10 points.
    Failure to hit target with second round, if fired _ _ _ _ _ _ _ _ _ _ _ _ _ _ _ _ _ 5 points.

f. See figure 57 for illustration of an appropriate score card.

131. Table VII—Moving Target Exercise (Shot)

a. The purpose of this exercise is to test the ability of the gunner to deliver effective fire on a moving target. The gunner fires at moving targets, using the burst-on-target method of adjustment for the second round. The gunner will be limited to two rounds of ammunition for each of the five targets he is to engage.

NAME _SMITH, JOHN V._
RANK _PFC_ SN _38132977_
DATE _4 FEB 54_

POSSIBLE SCORE ___100___
TOTAL CUTS ___28___
TOTAL SCORE ___72___

TANK GUNNERY QUALIFICATION COURSE SCORE CARD

TABLE VI (SERVICE FIRING, HE AND SHOT ADJUSTMENT).

			A	B		C	
TRIAL	NUMBER OF ROUNDS	POSSIBLE POINTS	TIME 1ST ROUND FIRED (Seconds)	TARGET HIT 1ST ROUND		TARGET HIT 2D ROUND	
				YES	NO	YES	NO
TARGET 1	2	25	12	X			
2	2	25	18	X			
3	2	25	11		X	X	
4	2	25	9		X	X	

Cuts:

	MAXIMUM CUTS	TARGETS				TOTAL CUTS
		1	2	3	4	
A. Failure to fire 1st round within 15 seconds	5 points	0	5	0	0	5
(Deduct one point for each second over 15.) Maximum cut.	5 points	0	3	0	0	3
B. Failure to hit target with 1st round.	10 points	0	0	10	10	20
C. Failure to hit target with 2d round.	5 points	0	0	0	0	0
(In case the target is hit by 1st round the 2d round will not be fired.)				TOTAL		28

Satisfactory Score -- 70 points.

Lt Leonidas V. James
Examining Officer's Signature

Figure 57. Score card for table VI.

b. The exercise is fired from a stationary tank at moving targets (6 x 6-foot panels) at unknown ranges, varying from 700 to 1,500 yards. Either a powered target or a towed target may be used. To vary the range from tank to target, either the tank or the targets may be moved to different locations. The target will be exposed for approximately 300 yards and will travel that distance at a constant speed of between 8 and 15 miles per hour.

c. The examiner will lay the gun for direction for each target while issuing a five-element initial fire command. He will index on the rangefinder the correct range to the target with the computer switch on. Time for each problem starts when the command FIRE is announced in the initial fire command.

d. The gunner will use the burst-on-target method of adjustment to fire the second round. This round will be fired even if the first round is a target hit.

e. The exercise will be scored as follows:

Table VII

Possible score: 100

Trial	Number rounds	Possible	Minus	Score
Target 1	2	20		
Target 2	2	20		
Target 3	2	20		
Target 4	2	20		
Target 5	2	20		

Cuts for each target:

Failure to fire first round within 15 seconds_____ 5 points.

Deduct 1 point for each additional second over 15 to fire first round, up to 20 seconds_____ 5 points.

Failure to hit target with first round_____ 5 points.

Failure to hit target with second round_____ 5 points.

f. See figure 58 for illustration of an appropriate score card.

132. Table VIII—Range Card Firing Exercise

a. The purpose of this exercise is to test the ability of the gunner to determine prearranged firing data for selected targets and to engage area type targets successfully with HE ammunition under conditions of restricted visibility, and to afford night firing practice.

b. Five 6 x 6-foot panels will be placed in a wide lateral area, at ranges varying from 800 yards minimum to 3,500 yards maximum, and at different angles of site. Panels will be numbered consecutively from left to right and will be visible from a suitable firing position.

c. Examining personnel will accurately compute the following data for each panel:

    (1) The azimuth indicator reading from an aiming stake or reference point.

    (2) The quadrant elevation, gun to target (elevation for range plus angle of site), with the elevation quadrant. This is determined as follows:

        (*a*) Check and adjust the elevation quadrant with the gunner's quadrant.

        (*b*) Index HE in the computer.

        (*c*) Determine the range to the target with the range finder.

NAME __SMITH, JOHN J.__

RANK __Pfc__ SN __38132977__

DATE __5 FEB 54__

POSSIBLE SCORE _____ 100

TOTAL CUTS _____ 17

TOTAL SCORE _____ 83

TANK GUNNERY QUALIFICATION COURSE SCORE CARD

TABLE VII (SERVICE FIRING MOVING TARGET).

TRIAL	NUMBER OF ROUNDS	POSSIBLE POINTS	A — TIME 1ST ROUND FIRED (Seconds)	B — TARGET HIT 1ST ROUND YES	B — TARGET HIT 1ST ROUND NO	C — TARGET HIT 2D ROUND YES	C — TARGET HIT 2D ROUND NO
TARGET 1	2	20	11		X	X	
2	2	20	12	X		X	
3	2	20	14	X		X	
4	2	20	17	X		X	
5	2	20	13		X	X	

Cuts:

	MAXIMUM CUTS	TARGETS 1	2	3	4	5	TOTAL CUTS
A. Failure to fire 1st round within 15 seconds	5 points	0	0	0	5	0	5
(Deduct one point for each second over 15.) Maximum cut.	5 points	0	0	0	2	0	2
B. Failure to hit target with 1st round	5 points	5	0	0	0	5	10
C. Failure to hit target with 2d round	5 points	0	0	0	0	0	0

TOTAL __17__

Satisfactory Score _____ 70 points.

Ct Leonidas K. Jones
Examining Officer's Signature

Figure 58. Score card for table VII.

(*d*) Lay the aiming point of the sight reticle on the center of the target, as would be done in direct fire.

(*e*) Without disturbing the lay of the gun, measure the existing elevation with the elevation quadrant.

> *Note.* If the reading on the micrometer dial is between markings, record to the next higher whole mil.

d. Ten E-type silhouette targets will be placed around the panel at which the gunner will fire. These panels will be placed as shown in figure 59.

e. The exercise will be conducted as follows:

(1) *Part I.* The gunner will be required to prepare a range card

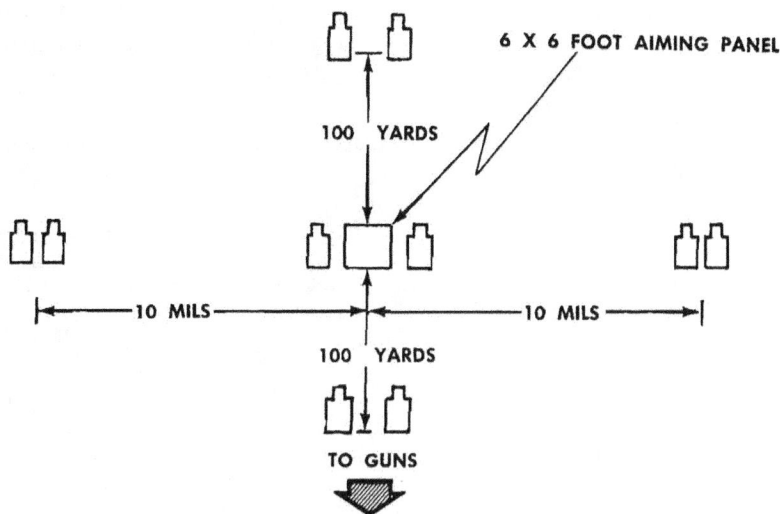

Figure 59. Range card firing exercise.

for the area, using the panels as likely targets. Information to be recorded on the range card will include:

(*a*) Aiming stake or reference point.

(*b*) Target (panel) numbers (left to right).

(*c*) Deflection (azimuth indicator reading) from aiming stake to each target.

(*d*) Range to each target in yards.

(*e*) Quadrant for HE ammunition.

(2) *Part II*. After the range card is prepared, the direct-fire sights will be covered (or the exercise may be fired at night) and the gunner will be required to fire on one of the panels, using his prearranged firing data. The examiner will, using the data computed by the gunner to one of the panels, issue an initial fire command; for example—

Element	*Example*
Alert	GUNNER.
Ammunition	HE.
Range	QUADRANT 13.
Direction	DEFLECTION 269 LEFT.
Description	CONCENTRATION 3.
Command to fire	FIRE.

The gunner will set off the data as announced in the initial fire command and fire only the first round. The gunner then will add 1 mil in elevation and simulate firing the second round; drop 2 mils and simulate firing the third round; add 1 mil, traverse right 10 mils, and simulate firing the fourth

149

round; traverse left 20 mils and simulate firing the fifth round. He will announce ON THE WAY as he simulates firing each round. Time will be recorded from the command FIRE.

Note. Elevation will be changed 1 mil to effect a 100-yard change.

f. The exercise will be scored as follows:

Table VIII

Possible score: 100

Part I_____ Total value 50 points.

For failure to obtain correct azimuth indicator reading within plus or minus 1 mil on each target, cut 5 points for each mil (maximum cut 25 points_____ 25 points.

For failure to obtain correct quadrant reading within plus or minus 1 mil on each target, cut 5 points for each mil (maximum cut 25 points)_____ 25 points.

Part II_____ Total value 50 points.

For each five seconds, or fraction thereof, that the time to fire the first round exceeds 25 seconds, cut 5 points (maximum cut 25 points)_____ 25 points.

For failure to hit in target area (20 mils by 200 yards), cut_____ 25 points.

g. See figure 60 for illustration of an appropriate score card.

Co _A_

Bn _12T_

NAME *SMITH JOHN J.*

RANK *PFC* SN *3813297*

DATE *6 FEB 54*

POSSIBLE SCORE	100
TOTAL CUTS	*15*
TOTAL SCORE	*85*

TANK GUNNERY QUALIFICATION COURSE SCORE CARD

TABLE VIII (RANGE CARD FIRING EXERCISE)

PART I		ROUNDS					TOTAL CUTS
		1	2	3	4	5	
(50 points)	1. Failure to obtain correct azimuth indicator reading within plus or minus 1 mil on each target. (Cut 5 points for each mil error not to exceed a total of 25 points.)	5	0	0	5	0	10
	2. Failure to obtain correct quadrant reading within plus or minus 1 mil on each target. (Cut 5 points for each mil error not to exceed a total of 25 points.)	0	5	0	0	0	5

PART II		TIME FIRST RD FIRED (SEC)	
(50 points)	1. Failure to fire first round within 25 seconds. (Cut 5 points for each five seconds or fraction thereof over 25 seconds not to exceed 25 points.)	19	0
	2. Failure to hit target area 20 mils by 200 yards. (Cut 25 points.)	HIT	0
	TOTAL CUTS		15

Satisfactory Score _____ 70 points.

Lt Leonidas K. James
Examining Officer's Signature

Figure 60. Score card for table VIII.

CHAPTER 6

INSPECTION AND MAINTENANCE

Section I. INTRODUCTION

133. General

The tank commander is responsible for insuring that required inspections are made. Mechanical efficiency is essential to the operation of tank units; therefore, each tank must be inspected systematically at intervals during each day of use. Defects thus can be discovered and corrected before they result in mechanical damage or failure. Crew members make their individual inspections and report the results to the tank commander, who, in his own report, lists all items requiring the services of maintenance personnel. In supervising driver maintenance or other services performed at periodic intervals and from day to day, the tank commander delegates responsibility to crew members as necessary. Maintenance procedures omitted from this manual are set forth in detail in TM 9–7012.

134. Maintenance To Be Performed

a. Inspections are made of all personnel equipment and weapons, fire-control equipment, communication equipment, vehicle equipment, and mechanical features of the vehicle. Inspections of instruments, radio sets, lights, tracks, suspension system, and engine performance are made in accordance with provisions of appropriate technical manuals. The driver fills in DD Form 110 (Vehicle and Equipment Operational Record), and the loader fills in DA Form 11–238 (Operator First Echelon Maintenance Check List for Signal Corps Equipment), indicating thereon deficiencies and maintenance work required. Any irregularities entered on these forms which are not repaired before the tank is again used will be re-entered continually until the deficiency has been corrected.

b. In succeeding paragraphs, the duties of the crew members are given in chart form to facilitate preliminary training of the soldier in performing crew maintenance. It is not intended that these procedures be followed as a precision exercise, such as in crew drill or gun drill; they are a guide for performing crew maintenance in a logical and efficient manner with a minimum loss of time.

Section II. CREW MAINTENANCE PROCEDURE

135. Before-Operation Service

This inspection begins with the tank locked and covered by a tarpaulin, with the gun in the traveling position. The turret is traversed as necessary to facilitate the various operations. A minimum of two manhours is required to perform this service properly.

Tank commander	Gunner	Driver	Loader
Command: FALL IN; PREPARE FOR INSPECTION.	Form in front of tank	Form in front of tank	Form in front of tank.
Inspect crew	Stand inspection	Stand inspection	Stand inspection.
Command: PERFORM BEFORE-OPERATION SERVICE.	Untie tarpaulin ropes	Fill out DD Form 110 during inspection.	Fill out DA Form 11–238 during inspection.
Supervise inspection and filling out of DD Form 110 and DA Form 11–238.			
Remove, fold, and stow tarpaulin	Assist tank commander	Inspect ground beneath tank for fuel and oil leaks. Inspect tracks and suspension.	Assist tank commander.
Supervise	Mount right front fender; open loader's hatch and enter tank; unlock tank commander's and driver's hatches.	Procure hand tools and lay out on left front fender.	Mount left front fender. Check contents of stowage boxes, pioneer tools, and tow cable.

153

Tank commander	Gunner	Driver	Loader
Supervise	Remove breech cover; inspect breech of 90-mm gun and replenisher indicator. Pass all covers to the loader; unlock turret lock. Check main and auxiliary engine air filters and oil level in hydraulic reservoir. Dismount.	Mount tank via left front fender and check fuel. Open engine compartment grilles. Check oil level in main and auxiliary engines and transmission for presence of oil.	Remove and stow muzzle cover. Unlock gun travel lock. Check stowage and lashing of personal gear. Receive and stow covers (breech and azimuth indicator); mount antenna. Assist driver.
Procure cleaning rods and material	Help swab bore of 90-mm gun and MGs. Take mounted post.	Help swab bore of 90-mm gun and MGs. Take mounted post.	Help swab bore of 90-mm gun and MGs. Take mounted post.
Check lights	Inspect sights and turret controls. Traverse turret to check azimuth indicator.	Check fixed fire extinguisher; close master switch; start auxiliary engine. Check operation of heater, steering controls, lights, horn, drain valves, and fuel cut-off.	Check portable fire extinguisher, operation of ventilator blower, water, first-aid kit, rations, 90-mm armament tools, spare parts, spare recoil oil, and engine oil.

Observe main engine exhaust; make visual check for oil leaks, vibration, and unusual noises.	Help attach shell bags; check bore and breech of tank gun.	Start main engine. Stop auxiliary engine. During warmup period check instruments, smoothness of operation, and any unusual noises.	Attach empty cartridge bag; mount coaxial MG and adjust headspace on MG; check mount and firing mechanism.
Check external interphone- - - - - - - - -	- - - - - - - - - - - - - - - - - -	Turn on external interphone. After warm-up period, check magneto operation.	Turn on radio; check with tank commander on external interphone.
Check transmission oil level; close grilles over engine compartment.	Check headspace on caliber .50 MG.	Check all publications pertinent to vehicle (TM, LO, Form SF-91, and DA Form 614).	Check hand grenades and caliber .45 ammunition (if applicable).
Direct driver to move tank forward two tank lengths. Inspect right and left track and suspension. Direct driver to move to original position. Check tools and stow.	Clean periscopes and vision blocks.	Take commands from tank commander.	Finish operator daily maintenance on radio equipment as prescribed in DA Form 11–238.
Take mounted post. Check: carbine and ammunition, power control, binoculars, flag set, and operation of radio.	Traverse turret to traveling position or firing position as directed.	Check: submachinegun and ammunition, periscopes, and escape hatch.	Lock gun travel lock if gun is put in travel position.
Connect breakaway plugs- - - - - - - - Check interphone system by commanding:	Connect breakaway plugs- - - -	Connect breakaway plugs- - - -	Connect breakaway plugs.

Tank commander	Gunner	Driver	Loader
REPORT------	Report: GUNNER READY------	Report: DRIVER READY------	Report: LOADER READY.
Report READY to platoon leader.			

136. During-Operation Service

This is a continuous process for all crew members during operation of the vehicle. All crew members must remain on the alert at all times for unusual noises and conditions, reporting to the tank commander if any are discovered.

Tank commander	Gunner	Driver	Loader
Check operation of radio and interphone system; observe security of antenna; check turret-mounted caliber .50 MG, and other visible outside equipment.	Check operation of sighting and fire-control equipment; check elevating and traversing mechanism if gun is in firing position.	Observe instruments; check operation of controls.	Check security of equipment in turret, coaxial machinegun, radio, portable fire extinguisher.

137. At-the-Halt Service

The length of time for the halt is the basis for determining how much of the following services will be completed; priority should be given to items accordingly. The tank commander will be informed of the time allotted for the halt and will indicate to his crew just how much of this time will be allotted to maintenance and inspection, and how much for relief of the crew members. As the result of training and experience, the crew will learn approximately what can be accomplished in a given length of time.

Tank commander	Gunner	Driver	Loader
Command: PERFORM AT-HALT SERVICE. (Use interphone system when applicable.) Disconnect breakaway plugs, supervise inspection.	Disconnect breakaway plugs. Release turret lock; check operation of turret traverse and gun elevation systems.	Disconnect breakaway plugs. Fill out DD Form 110 (during-operation and at-halt sections); idle main engine properly; check instruments.	Monitor radio and man turret-mounted MG (if applicable). If not applicable, check items listed under *loader* in before-operation section.
Dismount to rear deck; unlock gun travel lock (if applicable); open engine compartment doors, check transmission oil while engine is running; inspect operation of main engine and security of components.	Check sight adjustment, sighting and fire-control equipment, coaxial MG and mount, firing mechanisms, security of air cleaners.		
Clean outside of all turret periscopes and vision blocks.	------------------------	Stop main engine -------	Inspect communication equipment in turret.

Tank commander	Gunner	Driver	Loader
Inspect operation of auxiliary engine; check main engine oil level. Close engine compartment doors. Check pioneer tools and tow cable.		Start auxiliary engine. Stop auxiliary engine. Check driver's compartment for fuel and oil leaks; check fixed fire extinguishers, brakes, and controls.	
Lock gun in travel lock (if applicable).	Traverse turret to enable commander to lock gun in travel lock.		
Dismount; inspect track and suspension; check underneath vehicle for fuel or oil leaks; help driver check lights.	Check security of equipment stowed in turret.	Check service and blackout lights; clean periscope.	
Take mounted post	Take mounted post		Take own mounted post when commander mounts.
Command: REPORT	Report: GUNNER READY	Report: DRIVER READY	Report: LOADER READY.
Command: FALL OUT FOR BREAK (if applicable, alternate in manning of turret-mounted caliber .50 MG).			

158

Command: MOUNT (if dismounted).

Tank commander	Gunner	Driver	Loader
Connect breakaway plugs----------	Take mounted post; connect breakaway plugs.	Take mounted post; connect breakaway plugs.	Take mounted post; connect breakaway plugs.

138. After-Operation Service

Immediately after operation, the tank is given the service and maintenance necessary to prepare it in every way for sustained operation. This covers in practically the same order all points listed in the before-operation service. Obviously, more extensive servicing and maintenance are required. During this operation, the vehicle is cleaned, serviced, and replenished with fuel, oil, grease, ammunition, first-aid equipment, water, and rations. Refer to the vehicle lubrication order for the proper types and amount of oil and greases and intervals of use. All safety precautions against fire must be observed while refueling. A portable fire extinguisher must be available on the rear deck of the tank and must be manned by a crew member. Safety precautions must be carefully observed in handling ammunition.

Tank commander	Gunner	Driver	Loader
Command: PERFORM AFTER-OPERATION INSPECTION AND MAINTENANCE. Supervise operation.	Clean inside turret, tank gun, turret-mounted caliber .50 MG, and coaxial MG; replenish ammunition, water, rations, and first-aid equipment.	Fill out appropriate section of DD Form 110 during inspection. Idle engine properly before stopping; clean outside of vehicle; clean engine and engine compartment and driver's compartment; replenish fuel, oil, and grease.	Help gunner clean inside turret and turret-mounted and coaxial MG; replenish ammunition, water, rations, and first-aid equipment.
Check DD Form 110 and DA Form 11-238 for completeness; make final inspection of vehicle; report READY to platoon leader.		Complete DD Form 110 and give to tank commander.	Complete DA Form 11-238 and give to tank commander.

139. Weekly Preventive Maintenance Service

a. This maintenance service is performed weekly in addition to the daily maintenance services. It is also performed after each field operation in combat and on maneuvers. During garrison operation, allowances must be made for these services in preparing training schedules and work details. In combat and maneuvers, provisions must be made to allow time for crew members to perform this preventive maintenance. A minimum of 16 manhours is required to perform this service properly.

b. TM 9–7012 and LO 9–7012 must be complied with in performing this service. To facilitate preliminary training of the tank crew procedures in the following sequence are suggested.

Tank commander	Gunner	Driver	Loader
Command: FALL IN, PREPARE FOR INSPECTION.	Form in front of tank	Form in front of tank	Form in front of tank.
Inspect crew	Stand inspection	Stand inspection	Stand inspection.
Supervise inspection	Clean and paint any rusty spots in turret (when applicable).	Clean engines and engine compartment; make detailed inspection of main and auxiliary engines; service batteries.	Perform radio operation for weekly maintenance as prescribed by DA Form 11–238.
Assist driver; procure hand and pioneer tools, clean and check; clean and spot-paint rusty spots on outside of vehicle.	Inspect tracks and suspension system; tighten track wedge-bolts and center nuts as required.	Start main engine and drive vehicle forward as required. Clean and paint rusty spots in driver's compartment.	Assist gunner in inspection and service of tracks and suspension system.
Check DD Form 110 and DA Form 11–238 for completeness; make final inspection of vehicle; report READY to platoon leader.	Help lubricate	Lubricate as required; complete DD Form 110 and give to tank commander.	Help lubricate.

APPENDIX I

REFERENCES

Pam 310–series	Military publications.
Pam 108–1	Index of Army Motion Pictures, Television Recordings, and Film Strips.
SR 320–5–1	Dictionary of United States Army Terms.
SR 320–50–1	Authorized Abbreviations.
SR 385–310–1	Regulations for Firing Ammunition for Training, Target Practice, and Combat.
FM 5–25	Explosives and Demolitions.
FM 17–12	Tank Gunnery.
FM 21–5	Military Training.
FM 21–6	Techniques of Military Instruction.
FM 21–8	Military Training Aids.
FM 21–30	Military Symbols.
FM 21–60	Visual Signals.
FM 23–55	Browning Machinegun, Caliber .30.
FM 23–65	Browning Machinegun, Caliber .50.
TM 9–7012	90-mm Gun Tank, M48.
TM 9–1901	Artillery Ammunition.
TM 11–284	Radio Sets AN/GRC–3, –4, –5, –6, –7, and –8.
TM 11–704	Auxiliary Interphone Equipment, AN/VIA–1.

APPENDIX II

STOWAGE

Proper stowage of tank equipment is necessary for the efficient functioning of the tank and crew. First, each crew member must ascertain whether or not the equipment necessary to perform his duties is present. Second, and equally important, this equipment must be stowed in the proper place in order to be available when needed. So that these conditions can be met, a list of vehicle stowage and special tools has been prepared for the vehicle. The list of vehicle stowage and special tools designates an exact location either on or within the tank for every piece of authorized equipment, including personal equipment.

Name of part	Quantity required per vehicle	Where carried
ARMAMENT		
GUN, 90-mm, M41_____	1	On gun mount.
MOUNT, Combination Gun, 90-mm, T148.	1	Installed in turret.
GUN, Machine, Cal .50, Browning, M2, HB, flexible.	1	On mount on top of turret.
OR		
GUN, Machine, Cal .50, Browning, M2, HB, TT, w/o retracting slide (use w/MOUNT, A097–8683227).	1	
MOUNT, Machine gun, Cal .50_____	1	On commander's cupola.
OR		
MOUNT, Machinegun, TT, Cal .50____	1	
OR		
MOUNT, Machinegum, AA, Cal .50____	1	
GUN, Machine, Cal .30, Browning, M1919A4E1, Flexible.	1	Mounted on combination gun mount.
MOUNT, Tripod, Machinegun, Cal .30, M2.	1	Right fender box.
SIGHTING AND FIRE CONTROL		
COMPUTER, Ballistic, T31_____	1	Right turret wall.
OR		
COMPUTER, Ballistic, T30_____	1	
DRIVE, Ballistic, T24E2_____	1	Turret ceiling.

162

Name of part	Quantity required per vehicle	Where carried
SIGHTING AND FIRE CONTROL— Continued		
FINDER, Range, T46E1	1	Turret ceiling.
INDICATOR, Azimuth, T28	1	Right turret wall.
LIGHT, Instrument, M30	1	In turret above gun mount.
MOUNT, Periscope, T184	1	On ceiling.
MOUNT, Telescope, T191	1	Right coaxial mount.
PERISCOPE, M17 (1 spare)	5	Four installed around commander's cupola; one spare behind commander's seat.
PERISCOPE, M20	1	In Mount, Periscope, T184.
OR		
PERISCOPE, M20A1	1	
PERISCOPE, T36 (1 spare)	4	Three installed for driver; one spare right front of hull.
PERISCOPE, T41 (night vision PERISCOPE, to replace one DRIVER'S PERISCOPE).	1	On turret platform.
QUADRANT, Elevation, M13	1	On ballistic drive, T24E2.
QUADRANT, Gunner's, M1, w/ CASE, Carrying, M18.	1	In bracket, in turret.
OR		
QUADRANT, Gunner's, M1A1, w/ CASE, Carrying.	1	
SETTER, Fuze	1	In oddments tray.
TABLE, Firing, FT–90	1	In pamphlet bag.
TELESCOPE, T156E1	1	In Mount, Telescope, T191.
TOOLS AND EQUIPMENT FOR TANK, M48		
BAG, pamphlet, assy	1	In turret bustle behind commander.
BAG, tool, empty, 8½ in. high, 6 in. wide, 19½ in. lg.	1	Left rear fender box.
BAR, cross, socket wrench, rd solid, 7⁄16 in. diam, 8 in. lg.	1	In tool bag.
BAR, crow, pinch pt, 1¼ in. diam, 60 in. lg.	1	On left front fender box.
BAR, jimmy, stght, ½ in. blade width, 11⅞ in. lg.	1	Turret ceiling over radio.
BAR, socket wrench, extn, ½ in. sq-drive, 5 in. lg.	1	In tool bag.
BAR, socket wrench, extn, ½ in. sq-drive, 10 in. lg.	1	In tool bag.
BAR, socket wrench, extn, ¾ in. sq-drive, 8 in. lg.	1	In tool bag.
BAR, socket wrench, extn, ¾ in. sq-drive, 16 in. lg.	1	In tool bag.

163

Name of part	Quantity required per vehicle	Where carried
TOOLS AND EQUIPMENT FOR TANK, M48—Continued		
BASE, mounting, refueling pump, hand, assy.	1	In engine compartment.
BOX, grenade, assy_____	1	In rack behind commander.
BOX, spare bulbs_____	1	In rack behind commander.
CABLE, towing, 1⅛ in. wire rope diam, 10 ft lg (w/s EYES, 1½ x 3¼ in).	2	On rear hull plate.
CHISEL, machst, hand, cold, S, ¾ in. cut, 7 in. lg.	1	In tool bag.
CLIP, locking, S, ¼ in, diam wire, 1.250 pin groove, 3 in. over-all lgh.	1	Right fender box.
COVER, azimuth indicator, assy_____	1	On azimuth indicator.
COVER, canvas, 12 x 12 ft_____	1	On turret in bustle rack.
COVER, grille, front, center, assy____	1	Right rear fender.
COVER, grille, rear, assy_____	1	Right rear fender.
COVER, grille, side, left, assy_____	1	Right rear fender.
COVER, grille, side, right, assy_____	1	Right rear fender.
COVER, grille, side, right, front, assy_	1	Right rear fender.
DIAGRAM, strap location_____	1	In pamphlet bag.
EXTENSION (adapter), lubr gun (flex hose, sleeve type), 12 in. lg.	1	In tool bag.
FILE, AS, hand, sm cut, 10 in_____	1	In tool bag.
FILE, AS, three sq, sm cut, 6 in_____	1	In tool bag.
FIXTURE, track connecting and link pulling, L and RH (in pairs).	1	Right fender box.
GUN, lubr, hand lever operated, high pressure, 15 oz. cap.	1	In right fender box.
HAMMER, machst, ball peen, 2 lb___	1	In tool bag.
HANDLE, socket wrench, hinged, ½ in. sq-drive, 18 in. lg.	1	In tool bag.
HANDLE, socket wrench, rtc, rvrs, ½ in. sq-drive, 11 in. lg.	1	In tool bag.
HANDLE, socket wrench, "T" sliding, ½ in. sq-drive, 9 in. lg.	1	In tool bag.
HANDLE, socket wrench, "T" sliding, ¾ in. sq-drive, 17 in lg.	1	In tool bag.
HOOK, attaching, tow cable_____	1	Right fender box.
LIGHT, magneto, timing_____	1	In tool bag.
OILER, pump, S, bent spout, 1 pt cap_	1	In right fender box.
PADLOCK, keyed interchangeably, br body, 1¾ in., w/CLEVIS, set of 4 PADLOCKS and 6 KEYS (set).	1	Three on fender boxes, one on loader's hatch.
PIN, grooved, headless, S, cd/zn finish, 4⅝ in. nom lgh (use w/HOOK, G104-7068219).	1	Right fender box.
PLIERS, comb, slip jt, w/cutter, 8 in. lg.	1	In tool bag.

Name of part	Quantity required per vehicle	Where carried
TOOLS AND EQUIPMENT FOR TANK, M48—Continued		
PUMP, gasoline, portable, hand lever operated, w/suction and dispensing HOSES.	1	Pump in engine compartment. Hoses in right fender box.
SCREWDRIVER, comm, normal duty, 6 in. blade, $\frac{5}{16}$ in. tip, $11\frac{1}{2}$ in. lg.	1	In tool bag.
SCREWDRIVER, sp purpose, $1\frac{1}{2}$ inch blade, $\frac{5}{32}$ x 0.030 in. tip, 4 in. lg.	1	In tool bag.
SCREWDRIVER, machst, extra hv-duty, 5 in. blade, $\frac{1}{2}$ in. tip, $9\frac{1}{2}$ in. lg.	1	In tool bag.
TAPE, friction, general use, black, width $\frac{3}{4}$ in., 8 oz roll.	1	In tool bag.
WIRE, S, carbon, low annealed soft, black, diam 0.080 in. (10 ft).	bulk	In tool bag.
WRENCH, adj, sgle open end, $1\frac{5}{16}$ in. jaw opng, 8 in. lg.	1	In tool bag.
WRENCH, adj, sgle open end, $1\frac{5}{16}$ in. jaw opng, 12 in. lg.	1	In tool bag.
WRENCH, engrs, dble open end, 15 deg angle, alloy–S, $\frac{5}{16}$ and $\frac{3}{8}$ in. opngs.	1	In tool bag.
WRENCH, engrs, dble open end, 15 deg angle, alloy–S, $\frac{7}{16}$ and $\frac{1}{2}$ in. opngs.	1	In tool bag.
WRENCH, engrs, dble open end, 15 deg angle, spear-hd, alloy-S, $\frac{9}{16}$ and $1\frac{1}{16}$ in. opngs.	1	In tool bag.
WRENCH, engrs, dble open end, 15 deg angle, spear-hd, alloy-S, $\frac{5}{8}$ and $\frac{3}{4}$ in. opngs.	1	In tool bag.
WRENCH, set or cap screw (hollow-hd), hex, $\frac{1}{8}$ in. hex, $\frac{1}{4}$ in. set screw, No. 8 cap screw.	1	In tool bag.
WRENCH, set or cap screw (hollow-hd), hex, $\frac{5}{32}$ in. hex, $\frac{5}{16}$ in. set screw, No. 10 and 12 cap screw.	1	In tool bag.
WRENCH, set or cap screw (hollow-hd), hex, $\frac{3}{16}$ in. hex, $\frac{3}{8}$ in. set screw, $\frac{1}{4}$ in. cap screw.	1	In tool bag.
WRENCH, set or cap screw (hollow-hd), hex, $\frac{1}{4}$ in. hex, $\frac{1}{2}$ in. and $\frac{9}{16}$ in. set screw.	1	In tool bag.
WRENCH, set or cap screw (hollow-hd), hex, $\frac{5}{16}$ in. hex, $\frac{5}{8}$ in. set screw, $\frac{3}{8}$ in. and $\frac{7}{16}$ in. cap screw.	1	In tool bag.
WRENCH, set or cap screw (hollow-hd), hex, $\frac{3}{8}$ in. hex, $\frac{3}{4}$ in. set screw, $\frac{1}{2}$ in. and $\frac{9}{16}$ in. cap screw.	1	In tool bag.

Name of part	Quantity required per vehicle	Where carried
TOOLS AND EQUIPMENT FOR TANK, M48—Continued		
WRENCH, set or cap screw (hollow-hd), hex, ⅝ in. hex, 1¼ in. set screw, 1 in. cap screw.	1	In tool bag.
WRENCH, sgle, open-end, 15 deg angle, 3⅜₁₆ in. opng, 28⅛ in. lg.	1	Right fender box.
WRENCH, socket, ½ in. sq-drive, 8 pt, ⅜ in. opng.	1	In tool bag.
WRENCH, socket, ½ in. sq-drive, 12 pt, 7/16 in. opng.	1	In tool bag.
WRENCH, socket, ½ in. sq-drive, 12 pt, ½ in. opng.	1	In tool bag.
WRENCH, socket, ½ in. sq-drive, 12 pt, 9/16 in. opng.	1	In tool bag.
WRENCH, socket, ½ in. sq-drive, 12 pt, ⅝ in. opng.	1	In tool bag.
WRENCH, socket, ½ in. sq-drive, 12 pt, ¾ in. opng.	1	In tool bag.
WRENCH, socket, ½ in. sq-drive, 12 pt, ⅞ in. opng.	1	In tool bag.
WRENCH, socket, ½ in. sq-drive, 12 pt, 1⅛ in. opng.	1	In tool bag.
WRENCH, socket, ¾ in. sq-drive, 12 pt, 15/16 in. opng.	1	In tool bag.
WRENCH, socket, ¾ in. sq-drive, 12 pt, 1¼ in. opng (center guide bolt) (use w/BAR, 41–B–154).	1	In tool bag.
WRENCH, socket, ¾ in. sq-drive, 12 pt, 1 5/16 in. opng.	1	In tool bag.
WRENCH, socket, ¾ in. sq-drive, 12 pt, 1½ in. opng.	1	In tool bag.
WRENCH, track adjusting	1	Right fender box.
TOOLS AND EQUIPMENT FOR GUN, 90-MM, M41		
ADAPTER, bore brush and wiper ring.	1	Left rear fender box.
BOOK, record, weapon, parts I and II (one required and installed at proofing facility, with appropriate entries made in accordance with SR 750–115–20).	1	In pamphlet bag.
BRUSH, bore, 90-mm, M19	2	Left rear fender box.
COVER, breech, 90-mm gun, assy	1	On gun.
COVER, canvas, brush, bore, M518	2	On bore brush.
COVER, gun book, M539	1	In pamphlet bag.
COVER, muzzle, assy	1	On gun.

Name of part	Quantity required per vehicle	Where carried
TOOLS AND EQUIPMENT FOR GUN, 90-MM, M41—Continued		
EYE, lifting (breechblock removing)__	1	In roll, gun tools and equipment.
HEAD, rammer, unloading, M16 (diam 3.52 in., 10⅛ in. lg).	1	Left rear fender box.
OIL, hydraulic, petroleum base (MIL–L–5606) gal.	1	Right fender box.
RING, wiper_____	1	Left rear fender box.
ROD, push, assembling and disassembling shaft, diam ½ in., 6 in. lg, pt ⅛ in.	1	In roll, gun tools and equipment.
ROLL, gun tools and equipment, assy	1	Turret bustle, below radio.
ROPE, manila, stranded, 3-strand, 6 ft length w/tied ends for use w/ EYE, lifting.	1	In oddments tray.
STAFF SECTION, T3 (alum)_____	5	Left rear fender box.
TOOL, breechblock removing_____	1	In roll, gun tools and equipment.
TOOL, ramming and extracting_____	1	Under radio.
WRENCH, spnr, face, pin type, c to c of pins 2 in., diam of pin ¼ in., 6¾ in. lg.	1	In roll, gun tools and equipment.
TOOLS AND EQUIPMENT FOR Mount, COMBINATION GUN, T148		
BAG, empty, ctg, cal. .30 and .50 machinegun.	2	On T148 combination mount.
GUN, lubr, 8 oz cap_____	1	In roll, gun tools and equipment.
TOOLS AND EQUIPMENT FOR GUN, MACHINE, CAL .50, BROWNING, M2, HB (FLEXIBLE OR TURRET TYPE)		
BRUSH, cleaning, cal .50, M4_____	4	In roll, gun tools and equipment.
CASE, jointed cleaning rod and brush, cal .50, M15.	1	With roll, gun tools and equipment.
COVER, machinegun, cal .50 (for flexible gun only).	1	On gun.
COVER, spare bbl, cal .50_____	2	On spare barrel.
EXTRACTOR, ruptured ctg case, cal .50.	1	In oddments tray.
GAGE, headspace and timing, cal .50_	1	In roll, gun tools and equipment.
HIDER, flash (M2)_____	1	On cal .50 machinegun.
ROD, cleaning, jointed, cal .50, M7___	1	In case, M15.
WRENCH, muzzle gland and adj screw.	1	In roll, gun tools and equipment.

Name of part	Quantity required per vehicle	Where carried
TOOLS AND EQUIPMENT FOR GUN, MACHINE, CAL .30, BROWNING, M1919A4E1, FLEXIBLE		
BRUSH, cleaning, cal .30, M2_____	4	In roll, gun tools and equipment.
BRUSH, cleaning, chamber, M6 (bristle).	2	In roll, gun tools and equipment.
CASE, cleaning rod, cal .30, M1_____	1	With roll, gun tools and equipment.
COVER, spare bbl_____	2	On spare barrel.
EXTRACTOR, ruptured ctg, MK IV_	2	In oddments tray.
HIDER, flash, cal .30, M6, w/spare parts.	1	On gun.
ROD, cleaning, jointed, M1 (3 sections w/HDL).	1	In case, cleaning rod, M1.
WRENCH, comb, M6_____	1	In roll, gun tools and equipment.
TOOLS AND EQUIPMENT FOR MOUNT, TRIPOD, MACHINE-GUN, CAL .30, M2		
HOOD, tripod mount_____	1	On tripod mount.
TOOLS AND EQUIPMENT FOR CARBINE, CAL .30, M1		
CASE, ammo carrying_____	1	In rack behind commander.
TOOLS AND EQUIPMENT FOR GUN, SUBMACHINE, CAL .45, M3A1.		
CASE, ammo carrying_____	1	In rack behind commander.
TOOLS AND EQUIPMENT FOR SIGHTING AND FIRE CONTROL		
LIGHT, instrument, for az ind_____	1	Right turret wall.
LIGHT, instrument, M36, for tel mt_	1	On telescope mount.
FILLER, assy, for Periscope, M20___	1	In roll, gun tools and equipment.
LIGHT, instrument, M36 for M20____	1	Turret ceiling.
SPARE PARTS FOR TANK, M48.		
CONNECTOR, fire extinguisher, cylinder valve to nozzle line.	3	In tool bag.

168

Name of part	Quantity required per vehicle	Where carried
SPARE PARTS FOR TANK, M48— Continued		
FITTING, LUBRICATION, hyd, surface check, stght, ⅛ NPTF, short, male.	6	In tool bag.
LAMP, elec, incand, min, 3 v, sgle-tun-fil No. 323 for az ind.	3	In box 7021398.
LAMP, elec, incand, min, 24–28 v, sgle-tun-fil, No. 623 for dome and inst.	2	In box 7021398.
LAMP, elec, incand, min, 24–28 v, sgle-tun-fil, No. 1251 (B. O.).	4	In box 7021398.
NIPPLE, TUBE, refrigeration and marine br, ¾ in.	3	In tool bag.
PLUG, PIPE, automotive, hex-hd, ¼ in.	12	In tool bag.
PLUG, PIPE, automotive, sq-socket M1, 1 in. (filler, final drive).	2	In tool bag.
PLUG, PIPE, automotive, sq-hd, S, ⅜ in. (filler compensating idler).	1	In tool bag.
SPARE PARTS FOR CORD, LIGHT, EXTENSION		
LAMP, elec, incand, min, 24–28 v, sgle-tun-fil, No. 1683.	1	In cord, light extension.
SPARE PARTS FOR GUN, 90-MM, M41		
MECHANISM, percussion, assy_____	1	In roll, gun tools and equipment.
GASKET, cop, soft, 0.56 ID, 0.25 thk (recoil piston brg sleeve plug).	1	In roll, gun tools and equipment.
PLUG, drain and fill, S, sq-hd, ½–20NF-3 (recoil mechanism filler).	1	In roll, gun tools and equipment.
SPARE PARTS FOR GUN, MACHINE, CAL .50 BROWNING, M2, HB (FLEXIBLE OR TURRET TYPE)		
BARREL, assy_____	1	In right fender box.
PARTS, spare, w/BOX, Cal .50 machinegun, combat vehicle. Composed of: 1 BOLT, alternate feed, assy 1 BOX, spare parts (empty) 1 EXTENSION, firing pin, assy 1 EXTRACTOR, assy	1	In box in oddments tray.

Name of part	Quantity required per vehicle	Where carried
SPARE PARTS FOR GUN, MACHINE, CAL .50 BROWNING, M2, HB (FLEXIBLE OR TURRET TYPE)—Continued		
PARTS, spare, w/BOX—Continued Composed of—Continued		
1 LEVER, cocking		
1 LOCK, accelerator stop		
1 PIN, cocking lever, assy		
1 PIN, firing		
1 SEAR		
1 SLIDE, sear		
1 SPRING, sear		
1 STOP, accelerator		
1 SWITCH, bolt		
SPARE PARTS FOR MOUNT, MACHINEGUN, AA, CAL .50		
HANDLE, gun charging_ _ _ _ _ _ _ _ _ _ _ _	1	On caliber .50 mount.
SPARE PARTS FOR MOUNT, MACHINEGUN, TT, CAL .50		
BRACKET, assy (for gun)_ _ _ _ _ _ _ _ _ _	1	In roll, gun tools and equipment.
CABLE, electric, 21 in. lg_ _ _ _ _ _ _ _ _ _ _	2	In roll, gun tools and equipment.
CHARGER, assy_ _ _ _ _ _ _ _ _ _ _ _ _ _ _ _	1	In roll, gun tools and equipment.
CHUTE, ejection_ _ _ _ _ _ _ _ _ _ _ _ _ _ _	1	In roll, gun tools and equipment.
CHUTE, slip, assy_ _ _ _ _ _ _ _ _ _ _ _ _ _	1	In roll, gun tools and equipment.
CLAMP, hose, S, cd or zn-pltd, 4 in. (cover).	1	In roll, gun tools and equipment.
CLAMP, ring, S, 3½ ID, 4¼ OD (gun cover).	1	In roll, gun tools and equipment.
COVER, canvas (gun)_ _ _ _ _ _ _ _ _ _ _ _ _	1	On gun.
COVER and CORD, assy (shield)_ _ _ _	1	On gun.
NUT, jam, hex, lt, S, phos-ctd ¼-28UNF-2B.	2	In roll, gun tools and equipment.
SCREW, machine, fil-hd, drilled hd, dld–f/lkg–wire, No. 10 (0.190)–32NF–2A x 1¼.	3	In roll, gun tools and equipment.
SCREW MACHINE, flat-hd, S, phos-ctd, No. 10 (0.190)–32NF–2 x ½.	2	In roll, gun tools and equipment.

Name of part	Quantity required per vehicle	Where carried
SPARE PARTS FOR MOUNT, MACHINEGUN, TT, CAL .50—Con.		
SCREW, MACHINE, rd-hd, S, cd or zn-pltd, No. 10 (0.190)–32NF–2A x ⅜.	1	In roll, gun tools and equipment.
SOLENOID_____	1	In roll, gun tools and equipment.
STRAP, S, 0.1196 thk w/2 drilled countersunk holes.	1	In roll, gun tools and equipment.
STUD, bolt, S, 1.43 in. lg_____	1	In roll, gun tools and equipment.
SUPPORT, gun, assy (front)_____	1	In roll, gun tools and equipment.
TRIGGER, assy_____	1	In roll, gun tools and equipment.
WASHER, lock, split, extra hv, No. 10 screw size.	1	In roll, gun tools and equipment.
SPARE PARTS FOR GUN, MACHINE, CAL .30, BROWNING, M1919A4E1, FLEXIBLE		
BARREL, assy_____	2	In turret above radio.
PARTS, spare, w/BOX, cal .30 machinegun, combat vehicle. Composed of: 1 BOLT, assy 1 BOX, spare parts (empty) 1 EXTRACTOR, assy 1 LEVER, cocking 1 PIN, cocking lever 1 PIN, firing, assy 1 ROD, driving spring, assy 1 SEAR 1 SPRING, driving 1 SPRING, sear, assy 1 TRIGGER	1	In box in oddments tray.
SPARE PARTS FOR SIGHTING AND FIRE CONTROL		
LAMP, elec, incand, min, 24–28 v, No. 313, for computer.	2	In spare bulb box.
LAMP, elec, incand, min, 2–3 v, No. 43, for ballistic drive.	2	In spare bulb box.
LAMP, elec, incand, min, 24–28 v, sgle-tun-fil, No. 313, for range finder.	3	In spare bulb box.
LAMP, elec, 24–28 v, 21 cp, sgle-bayonet base, GE No. 1203–S–8 bulb, for range finder.	7	In spare bulb box.

Name of part	Quantity required per vehicle	Where carried
SPARE PARTS FOR SIGHTING AND FIRE CONTROL—Con.		
LAMP, elec, incand, min, 3 v, No. 323, 0.19 amp, for az ind.	4	In spare bulb box.
LAMP, elec, incand, min, 3 v, sgle-tun-fil, No. 1325, for M36 light.	2	In spare bulb box.
LAMP, elec, incand, min, 24–28 v, sgle-tun-fil, No. 313, for T184 gun mount.	2	In spare bulb box.
LAMP, elec, incand, min, 24–28 v, sgle-tun-fil, No. 313, for tel mt.	2	In spare bulb box.
HEAD, assy for Periscope, T41_____	1	On turret platform.
SEAL for Periscope, M36_____	1	In box with head assembly.
HEAD, assy for Periscope, M20_____	1	In box below gunner's seat.
EQUIPMENT ISSUED BY OTHER TECHNICAL SERVICES		
AXE, hdl, chopping, sgle bit, std grade, wt 4 lb.	1	In rack, on left fender.
BRUSH, sash tool oval, 1⁷⁄₁₆ x 1¹⁄₁₆ x 2¾ inches.	2	Left rear fender box.
CAN, water, cap 5 gal_____	2	Outside turret.
CARRIER, wire cutter, M1938_____	1	In tool bag.
CORD, light extn, inspection_____	1	In tool bag.
CUTTER, wire, M1938_____	1	In tool bag.
EXTINGUISHER, fire, carbon dioxide, shatterproof, 5 lb, permanent shutoff, hand.	1	On turret platform.
FLAG SET, M238_____ Composed of: 1 CASE, CS–90 1 FLAG, MC–273 1 FLAG, MC–274 1 FLAG, MC–275 3 FLAGSTAFF, MC–270	1	In right side of turret by tank commander.
FLASHLIGHT, elec, hand, 2 cell_____	3	One in driver's compartment, two in turret.
FORM (envelope), DA 478_____	1	In pamphlet bag.
HANDLE, mattock, 36 in. lg_____	1	In rack on left fender.
KIT, first aid, motor vehicle, 12 unit__	1	In fender box or below radio.
MANUAL, supply, ORD 7 SNL G–254.	1	In pamphlet bag.
MANUAL, Technical, 9–7012_____	1	In pamphlet bag.
MATTOCK, pick, w/o hdl, wt 5 lb___	1	In rack on left fender.
MITTEN, asbestos, M1942, pair_____	1	In oddment tray.
ORDER, Lubr, 9–7012_____	1	In pamphlet bag.
PANEL MARKER, signal to aircraft, nylon VS17G VX, 17 in. lg, 24 in. wide.	2	Left front fender box.

172

Name of part	Quantity required per vehicle	Where carried
EQUIPMENT ISSUED BY OTHER TECHNICAL SERVICES—Con.		
RADIO SETS AND COMBINATIONS: (AN/GRC-3, -4, -7, or -8 or AN/VRC-7) (AN/GRC-3 or -7 and AN/ARC-3, or AN/ARC-27). (AUXILIARY INTERPHONE EQUIPMENT AN/VIA-1).		In SCR mounts.
SHOVEL, GP, "D" hdl, rd-pt........	1	In rack on left fender.
SLEDGE, blacksmith's dble face, 10 lb.	1	On left fender.
STOVE, gasoline burner, M42 w/ CASE, fuel capacity ½ pt.	2	Left front fender box.
TUBE, flex nozzle, cam type..........	2	In right fender box.
SPARE PARTS FOR TANK, M48		
SHOE, track assy..................	2	Right and left front fenders.
ARMAMENT		
CARBINE, Cal. 30, M1.............	1	Inside turret above radio.
GUN, Submachine, Cal .45, M3A1....	1	Left turret wall, above loader.
SIGHTING AND FIRE CONTROL		
BINOCULAR, M17A1....	1	In rack, behind commander.
AMMUNITION		
CARTRIDGE, Cal .30, for Carbine, Cal .30, M1.	180	In CASE, Ammunition D7052438 (in turret bulge).
CARTRIDGE, Cal .30, for gun, Machine, Cal .30, M1919A4E1.	5,900	2,200 in turret wall box; 2,200 in floor box (front); 500 each in three floor boxes under commander.
CARTRIDGE, Cal .45, for Gun, Submachine, Cal .45, M3A1.	180	In CASE, Ammunition D7052438 (in turret bulge).
CARTRIDGE, Cal .50, for Gun, Machine, Cal .50, Browning M2, HB.	525	One box (105) on gun mount; two boxes (210 each) on floor.
GRENADE, hand...................	8	In two boxes, D7388581.
ROUND, for Gun, 90-mm, M41.......	60	30 in turret; 30 in hull.

Name of part	Quantity required per vehicle	Where carried
EQUIPMENT ISSUED BY OTHER TECHNICAL SERVICES		
BATTERY, dry cell, 1½ v, 1¼ in. diam x 1¼ inches lg, BA–30.	14	Six in flashlights; eight in instrument lights.
CANTEEN, M1910, w/CUP AND COVER.	4	One in hull; three in turret.
PACK, field, cargo and combat, M1945.	4	Outside turret bustle.
SUSPENDERS, pack, field, cargo and combat.	4	Outside turret bustle.
RATIONS, individual combat_____	12	One in driver's compartment; eight in left front fender box; three in turret.

INDEX

[AG 470.8 (3 Aug 55)]

By order of the Secretary of the Army:

MAXWELL D. TAYLOR,
General, United States Army,
Chief of Staff.

Official:
JOHN A. KLEIN,
Major General, United States Army,
The Adjutant General.

Distribution:
Active Army:

CNGB (1)	Armd Div (25)
Tec Svc, DA (1)	Abn Div (5)
Tec Svc Ed (2)	Brig (2)
Hq CONARC (40)	Inf Regt (2)
Army AA Comd (2)	Armd Gp (5)
OS Maj Comd (5)	Armd Bn (5)
OS Base Comd (2)	Armd Co (5)
Log Comd (2)	USMA (5)
MDW (1)	Inf Sch (50)
Armies (10)	PMST (2)
Corps (5)	Mil Dist (2)
Inf Div (5)	

NG: State AG (6) ; units—same as Active Army except allowance is one copy to each unit.

USAR: Same as Active Army except allowance is one copy to each unit.

For explanation of abbreviations used see SR 320-50-1.

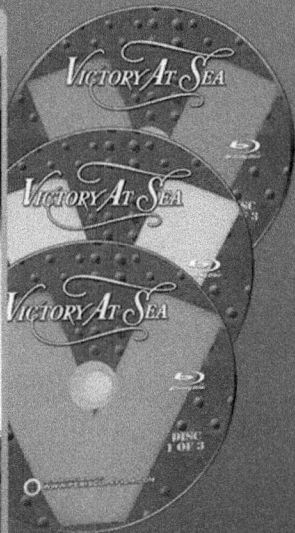

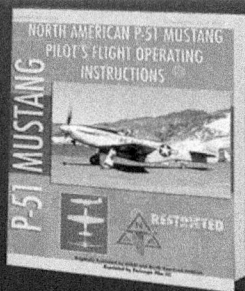

www.ingramcontent.com/pod-product-compliance
Lightning Source LLC
Chambersburg PA
CBHW052002090426
42741CB00008B/1505